PYROCENE PARK

PYROCENE PARK

A JOURNEY INTO THE FIRE HISTORY OF YOSEMITE NATIONAL PARK

STEPHEN J. PYNE

THE UNIVERSITY OF ARIZONA PRESS
TUCSON

The University of Arizona Press
www.uapress.arizona.edu

We respectfully acknowledge the University of Arizona is on the land and territories of Indigenous peoples. Today, Arizona is home to twenty-two federally recognized tribes, with Tucson being home to the O'odham and the Yaqui. Committed to diversity and inclusion, the University strives to build sustainable relationships with sovereign Native Nations and Indigenous communities through education offerings, partnerships, and community service.

ISBN-13: 978-0-8165-4923-8 (paperback)
ISBN-13: 978-0-8165-4924-5 (ebook)

Cover design by Leigh McDonald
Cover photo: *Glacier Point Yosemite Fire* by Darvin Atkeson – Yosemite Landscapes.com
Typeset by Sara Thaxton in 10/14 Warnock Pro with Acumin Variable Concept and Interstate Compressed

Library of Congress Cataloging-in-Publication Data
Names: Pyne, Stephen J., 1949– author.
Title: Pyrocene park : a journey into the fire history of Yosemite National Park / Stephen J. Pyne.
Description: Tucson : University of Arizona, 2023. | Includes index.
Identifiers: LCCN 2022031449 (print) | LCCN 2022031450 (ebook) | ISBN 9780816549238 (paperback) | ISBN 9780816549245 (ebook)
Subjects: LCSH: Wildfires—California—Yosemite National Park—History. | Forest fires—California—Yosemite National Park—History. | Fire management—California—Yosemite National Park—History.
Classification: LCC SD421.32.C2 P96 2023 (print) | LCC SD421.32.C2 (ebook) | DDC 363.37/90979447—dc23/eng/20220829
LC record available at https://lccn.loc.gov/2022031449
LC ebook record available at https://lccn.loc.gov/2022031450

Printed in the United States of America
♾ This paper meets the requirements of ANSI/NISO Z39.48-1992 (Permanence of Paper).

TO SONJA

OLD FLAME, ETERNAL FLAME

CONTENTS

ILLUSTRATIONS

Figures

Maps

PYROCENE PARK

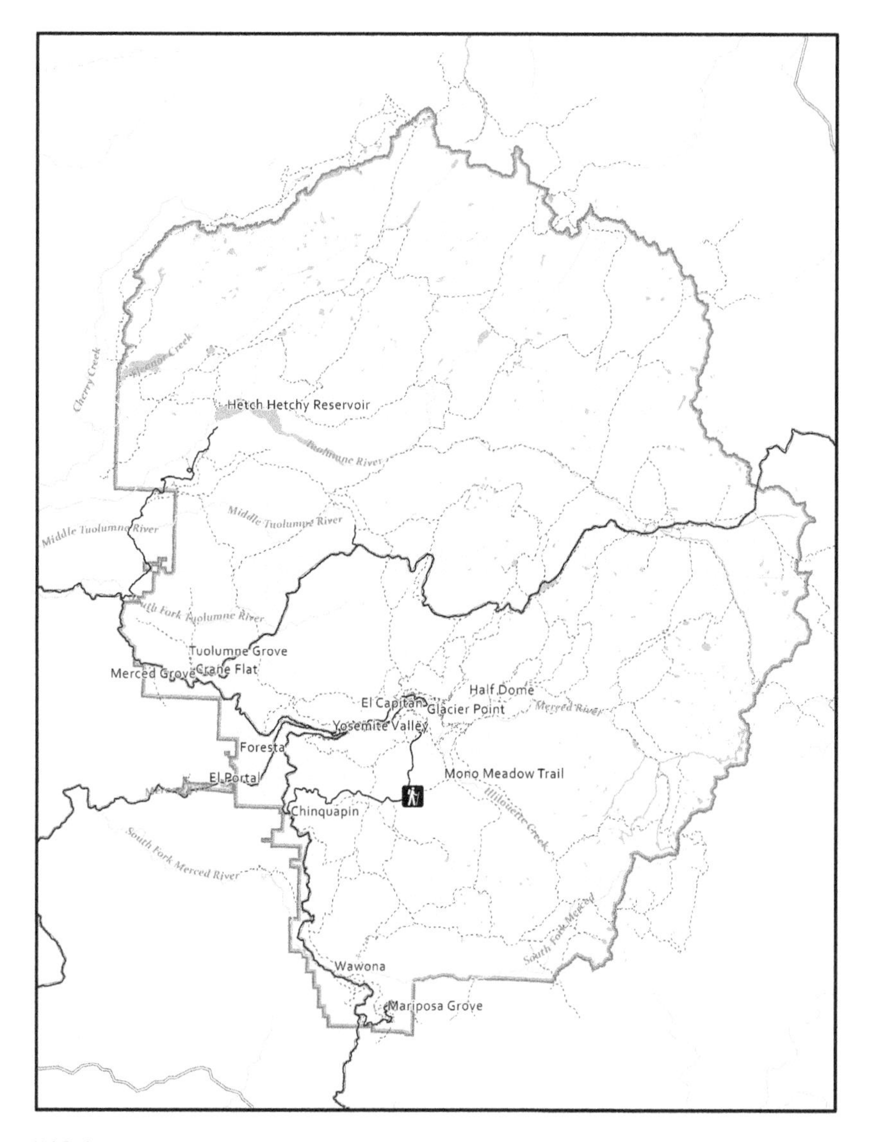

MAP 1 Yosemite National Park.

Prologue
Ice and Fire

A trek to the Illilouette began as a thought by Jan van Wagtendonk, evolved into a resolve by the park's upper administration, advanced to a project under the fire management program, and became a reality on September 13–15, 2021.

Behind that undertaking lay the massif of the Sierra Nevada Range, California's Mediterranean climate, a biota built to burn, humanity's monopoly over fire, America's halting history from laissez-faire burning to universal suppression to restoring good fires, Yosemite's status as an emblem of the wild, the Earth's hastening spiral from ice to fire, and those ineffable moments when planet and people converge.

• • •

The Illilouette Valley—hidden in the aesthetic shadow of Half Dome—is not a destination landscape. No John Muir has rhapsodized over its wild splendor. No Ansel Adams has immortalized it in photographs. No guidebooks identify it as one of Yosemite's many iconic scenes. It boasts no towering granite domes, no Big Trees, no historical markers, no cult of climbing routes. In a place that overflows with the photogenic and the monumental, it projects no special vision or public voice. It is neither in Yosemite Valley nor along the Range of Light that forms the Sierra Crest. Its trees are Jeffrey pine, lodgepole, and aspen patches, not giant sequoias.

Which makes all the more astonishing that the superintendent, deputy superintendent, chief ranger, wilderness policy and recreation planner, chief of resources management and science, chief of ecological restoration, vegetation ecologist, fire ecologist, wilderness manager, park physical scientist, chief of staff, fire management officer, deputy fire management officer, and fuels battalion chief—most of the governing cadre of the park concerned with Yosemite's natural endowment—along with two academics planned a three-day trek to the basin on September 13–15, 2021. These are the people who must decide how to manage the park's natural estate.

That domain has been undergoing a slow, now quickening upheaval that makes Yosemite a microcosm of the Earth. Nearly all Yosemite's fabled sites were shaped by Pleistocene ice as the planet flickered over the past 2.6 million years into and out of long glacial epochs broken by short bouts of warming. That ice was the most visible feature of a makeover that repeatedly recast the Earth's lands, seas, and air. At Yosemite it widened and deepened valleys, rounded exposed granite, cached moraine and soils, and scoured routes for runoff that became rivers and waterfalls. Over and again, the ice made its mark, departed, and repeated.

The last interglacial, known as the Holocene, began roughly 12,000 years ago. But something new intervened in the rhythm of returning ice. This time a fire-wielding creature, *Homo sapiens*, interacted with a progressively fire-receptive world. The cooling stalled, then reversed. It was as though the expected ice age had refracted through a pyric prism and re-emerged as a fire age. Fire replaced ice, fire drove off ice. Visible flames reshaped living landscapes of conifers, shrubs, grasses, and peat, while combustion hidden in machines, burning the fossilized residue of formerly living biomes—call them lithic landscapes—began reforging how humans lived on the land. When the effluent from that industrial-scale firing marinated the atmosphere with greenhouse gases, it perturbed the climate, which reconfigured everything it touched. Local fires massed into a globalized fire age.

Even Yosemite, a monument to ice, is being refashioned by the hastening fires. That is what makes the Illilouette, otherwise so mundane, of interest to park management: it is a place informed by fire. It is where the park sought to test the notion, an amalgam of hope and alarm, that good fires might restore the lost fires and help stave off the bad burns, the feral flames, and the megafires that a blowup fire age threatens. It is where a landscape bequeathed by the Pleistocene has morphed into a Pyrocene.

CHAPTER 1

Valley and Grove

The preliminaries begin by dropping off gear at the Park Service's Yosemite Valley corrals on the evening of September 11. Some of the routes to the valley follow old trails, etched by thousands of years' foot traffic before Europeans arrived. Most were constructed to bring the newcomers from Stockton and San Francisco to the scenic marvels. In 1856 the Mann Toll Horse Trail, later upgraded for stagecoaches, carried tourists the 45 miles between Yosemite Valley and Wawona, site of the Mariposa Grove of giant sequoias. Stage lines and horse trails were later paved for automobiles.

Packers will haul the duffels, camp gear, and stock the next morning to the Mono Meadow Trailhead, and there load for the haul in and out. The trek will consist of 15 people, the maximum allowed in a backcountry group.

• • •

The prehistory of fire is less easily assembled and cinched with diamond hitches. Little of the Sierra Nevada is not prone to burn. The western foothills of grass and oak burn. The chaparral shrublands burn. The montane forests and grassy valleys burn. The upland forests and subalpine meadows burn. They burn at different frequencies and scales—the lower grasses may burn annually and broadly, the subalpine woods spastically on an order of decades. Fire scars are as common and diagnostic in the Sierra biota as glacial gouging on its granite.

Lightning is abundant, which assures that with or without people the landscape will burn. But people have been more abundant, and they have added their fires to those kindled by lightning, although in many cases their fires competed with nature's because at any time and place, only one fire could burn. The peoples around Yosemite—the Miwok generally, the Ahwahneechee in the valley—used their tame fires to replace nature's wild ones. They burned routinely in Yosemite Valley, in the sequoia groves, across the landscape webbed with trails and patches useful for foraging and hunting. They made a second nature out of first nature's raw materials.[1]

If Yosemite records in miniature the story of the Pleistocene ice ages, the Miwok concentrated the story of a unique Pleistocene species, humans, who could claim a monopoly over the manipulation of fire. They did what no other creature could. With fire they made their world more habitable to themselves and those plants and animals they relied on.

• • •

Yosemite was America's first experiment with national parks. In 1864 Congress set aside Yosemite Valley and the Mariposa Grove of sequoias near Wawona from the public domain and delegated to California actual administration over them as a park. The state accepted the Yosemite Grant in 1866 and organized a commission of luminaries including Josiah Dwight Whitney, then director of the California Geological Survey, and Frederick Law Olmsted, a landscape architect famous for New York City's Central Park. The commission granted a local tour guide, Galen Clark, official standing as Guardian of the Yosemite for his self-proclaimed work as "protector."

Everyone who first encountered the valley noted the openness of its panorama, and all attributed that fact to the annual burning by the Ahwahneechees. The natives had ample reasons, both in the valley proper and across the surrounding countryside: they burned to keep grasslands open and attract game animals like deer to fresh forage; to keep trees from crowding out annual-grass prairies; to help harvest acorns; to stimulate willows useful for basketry; to promote berries; to open woods to assist hunting; to prevent shrubs and trees from sealing off corridors of travel or overrunning villages; to dampen a buildup of surface combustibles that could, under California's Mediterranean climate and autumn winds, explode violently. Signal fires passed messages by flame at night and smoke by day. Much of their fires were, as these rationales suggest, deliberate and purposeful, even artful. Tribes moved seasonally around the countryside, adjusting the timing of their fires with what the landscape offered for food and provisions. Written accounts testify to near-annual burning in Yosemite Valley. A reconstruction of presettlement fire regimes based on fire-scarred trees in the lower Tuolumne Valley suggests that fire returned, on average, about every 10 years. Other fires from accident, carelessness, untended campfires, or hostile flame accompanied those seasonal travels (accounts of abandoned campfires are common). Yet so extensive was fire that even fire littering or arson had to burn within the limits of previous burns. Past fires checked future fires.[2]

When Galen Clark initially visited the valley in 1855, it was wide-vistaed, with "no undergrowth of young trees to obstruct clear open views in any part of the valley." Looking back half a century later, he wrote that "a great change has taken place in Yosemite Valley since it was taken from the control of the native Indians who formerly lived there." In the early years, he insisted, when first visited by white people, three-fourths of the valley was open ground—meadows with grasses and flowering plants waist high. On the drier parts were scattered forest trees—pines, incense cedars, and black oaks—too widely separated to be called groves, leaving an unbroken, extensive vision up and down and across the valley from wall to wall on either side. He explained the why of the burning, and what happened when it ceased:

> The Indians kept the valley clear of thickets of young trees and brushwood shrubbery, so they could not be waylaid, ambushed, or surprised by enemies from outside, and to not afford hiding places for bears or undesirable predatory animals, and also to have clear ground for gathering acorns, which constituted one of their main articles of food. At the present time there is not more than one-fourth of the floor of the valley clear, open ground, as there was fifty years ago. Nearly all the open ground between the large scattering trees is now covered with a dense growth of young trees, which also extend out over hundreds of acres of the driest portion of meadow land. Every pine tree of the valley less than seventy-five feet high has grown from seed within the past fifty years.[3]

Clearly, for the Miwok, fire was not just a natural phenomenon like spring runoff or a simple tool like an arrowhead, but the domestication of a full-spectrum ecological process. It was a relationship.

• • •

Early settlement broke the sweep of burning by plowing, dismantling brush dams that helped keep meadows wet, planting orchards and fields, and loosing livestock. Ruptured dams promoted more dry sites in which trees could root. Apple trees did not underburn like black oaks. Roads and plows shattered the sweep of flame. Cattle, horses, and sheep metabolically slow-combusted what fire fast-combusted. Still, at the time of the 1864 cession, such intrusions were little more than vignettes on the landscape. Fire continued, the vista remained untrammeled. In the backcountry, burning by shepherds replaced that lost by the removal of the Miwok, with old burns still buffering new ones.[4]

The commissioners accepted as a primary duty that the valley and grove, along with "their surroundings," should be secured "from devastation by fire." "The care of the Guardian has prevented fires from running in the Big Tree Grove, and to a considerable extent has protected the valley from wanton injury." But to anyone like

Frederick Law Olmsted who regarded the "duty of preservation" as primary, fires were a simple act of vandalism, and worse, an enabler of more powerful intrusions. In his 1865 report to his fellow commissioners, he denounced the "Indians and others" who have set fires and killed trees, and noted that even the "giant tree" in the Mariposa Grove (and "probably the noblest tree now standing on the earth") had been injured—"burned completely through the bark near the ground for a distance of more than one hundred feet of its circumference." Preservation meant preservation from human finagling, of which fire was both visible and notably egregious. To the commission's proposal to lay out a road to the grove, Olmsted urged that a road completely encircle the Big Trees and so prevent fires "which nearly every year sweep upon it from the adjoining country, and which during the last year alone have caused injuries."[5]

What happened instead was a case-study in unintended consequences. The valley began silting in with trees, and the grove filled with white fir and incense cedar. The views—and a principal purpose of the park was to ensure those views in perpetuity—became obscured by trees. They blocked the grandest perspective on waterfalls, they cluttered the valley, they impeded movement. Meanwhile, visitors found that it was becoming impossible to see the Big Trees for the forest that enveloped them. The change became apparent in a handful of years. The same cause underwrote both landscapes: the absence of fire.

In 1889, before he resigned as a commissioner, William H. Mills confirmed not only the breakneck pace of forest encroachment in valley and grove but how native burning had routinely cleaned out the scenes before settlement. ("I have always respected the ability of the Indians to manage that valley.") At issue was not just a matter of appearance and open vistas: the amassing of combustibles threatened to stoke uncontrollable fires that could destroy both valley and grove. People would have to intervene, and personally he considered burning "a very good method of management"—certainly less intrusive than logging off the unwanted woods. The grove was in worse shape because the former fires, which "did no harm," had actually assured the survival of the Big Trees. Now, in fire's absence, wood was piling up around their bases "to the depth of four or five feet," and "in some dry season" the grove would burn, not benignly but ruinously. Custodian Clark had already begun quietly clearing in the grove.[6]

Yet from the onset of the park, a paradox emerged. Most close-observing settlers, naturalists, and commissioners understood the value of the annual burning conducted by the Miwok yet sought to suppress it or any avatars that might reincarnate at the hands of newcomers. Emil Ernst, long-serving park forester, noted that "up until about 1906 this policy of fire suppression was openly and actively condemned by the highest responsible officials who, however, disregarded their own opinions and carried out the fire suppression and fire prevention policies with which they heartily and honestly disagreed."[7]

Why? Ernst speculated that the program actually targeted the cattlemen and shepherds whose use of fire made their trespass possible, burning that officials regarded as promiscuous and, despite whatever ecological good might result, threatened social order and political authority. Denying them fire was a way to control their trespass. Abolishing the annual burning of the valley was collateral damage that seemed minor compared with the greater harm done by unrestricted grazing. Until it wasn't. Once established the policy persisted, and the cognitive dissonance it embodied became institutionalized. It continues today.

• • •

As the valley's human habitation shifted, so did the biota that sustained it. The purposes of the park shifted from gathering acorns to viewing waterfalls: the meadows mattered more for their scenic views than their ability to feed. The human history of the valley reconstituted itself accordingly. Not for the last time, Yosemite found itself between two fires.

One fire, and the only fire officially sanctioned in the valley, it seemed, was the Glacier Point Firefall devised by James McCauley, owner of the Four-Mile Trail from the valley to Glacier Point. On the Fourth of July in the early 1870s he burned a large bonfire at the point, then pushed the charred and glowing residue over the cliff. Visitors stared up amid the deepening dusk while embers rained down in fiery facsimile to the valley's waterfalls. Recreational firefalls to entertain visitors were acceptable; free-ranging fires through valley and grove were not.[8]

The second fire was feral. State administration was feeble and fraught, endlessly troubled by the dual imperatives to promote visitor use while simultaneously preserving the natural scene. The commission appeared inadequate to meet Yosemite's looming problems, not least the burgeoning tourism and facilities created to boost it further. Paradoxically, in removing the tamed burning of the Miwok, the park had made itself vulnerable to wild fire.

In 1889, after 25 years of state protection, the crisis the commission confronted was less with missing fires than all-too-real ones. In the summer of 1889 "that most despicable of crimes, forest arson," the commissioners thundered, started either by careless campers or conniving shepherds, turned the forest around the grove into a "flood of fire," a "conflagration" that "at times almost surrounded the Great Sequoia Grove and invaded it at many points." The fire highlighted what many critics regarded as a commission staffed by an inept, self-serving elite. The fire graphically symbolized the commission's failure.[9]

It was a year of fires throughout the West, in some ways foreshadowing the more potent Great Fires that blew up in 1910 and energized the young U.S. Forest Service (USFS). The threat to the Sierra sequoias galvanized legislation the next year to create Sequoia and General Grant National Parks, and to reconfigure the boundaries of Yosemite National Park, an arrangement accepted by California the next year. So while the state continued to manage the park until 1906, it was lost on no one that the failure to contain wildfire was the immediate provocation for the transfers. The urgency continued into 1891 as the Forest Reserve Act allowed the president to set aside public land as protected forests. The earliest decrees bolstered Yellowstone National Park. The next targeted the protected parks in the Sierra Nevada, along with mountain watersheds around the Los Angeles Basin. All accepted fire protection as a foundational charge.

That implied mandate acquired legal standing as Secretary of the Interior John Noble promulgated regulations that made it illegal to "start or kindle or allow to be started or kindled any fire in grass, leaves, underbrush, debris or dead timber down or standing." Granted a country full of flame, some from clearing, most from agriculture, some bad, most good, the blanket prohibition seems eccentric if not irrelevant. But the western forests and parks were not the primary inspiration: the Northeast was, notably the Adirondacks, which had suffered decades of blowups, with the worst yet to come. The man who prepared a management plan for the Adirondacks park, Charles Sargent, a Harvard professor and overseer of the Arnold Arboretum, had also conducted a survey of forest fires for the 1880 census, and later chaired the 1896 National Academy of Sciences' Forest Commission to prepare an organic act for the new forest reserves. Brash new foresters like Gifford Pinchot, a Pennsylvania native, projected the fire-catalyzed havoc among the logged landscapes of the Northeast westward. The easterner who went west was the vector for fire abolitionism.[10]

Solving the East's fire problem only made the West's worse. The rules criminalized burning that had gone on for millennia; penalties included liability for any damages that resulted. The rules were interpreted as an indirect way to rein in homesteading squatters, poaching loggers, and slovenly sheepmen. But by making no distinctions among fires, the prohibition anointed fire's removal as a default policy, and it confirmed the mismatch between what officials saw and what they said. It only remained to find a corps capable of applying that policy, that would be undaunted by fights against

FIGURE 1 Troop D, Ninth Cavalry, on the trunk of the Fallen Monarch, Mariposa Grove, 1906.

threats domestic and foreign. In 1891 the U.S. Cavalry rode into Yosemite, General Grant, and Sequoia National Parks, as it had into Yellowstone five years earlier.

The army soon found itself in a firefight.

CHAPTER 2

Fire and Sword

The journey to the trailhead follows the old Yosemite stage line and toll road that ran from the valley to the grove, the two anchor sites of Yosemite as a public park. The commission focused mostly on the valley. The army camped by the grove. But the expanded dimensions of the park and the need to control trespass, particularly by sheep herders, forced the cavalry to explore old trails and new sheep runs. Some would survive as routes; others would succumb to the abrasions of nature.

As the park evolved, it developed new cartographies of travel suited to the announced needs of the day. The earliest routes were seasonal songlines of hunting and foraging, and when connected, supported trade from the Great Basin to the Pacific Coast. The park reconfigured that geography to suit tourists; the army, to patrol for intruders. The Illilouette trekkers would begin with the old stage line, and then, as the army did in its time, and the Park Service later amplified, branch off to pursue its own mission.

• • •

"My duties were entirely new to me," wrote Captain Abram E. Wood of the Fourth Cavalry in his first annual report, "and I had no idea of what they were." So began the army's administration as it sent troops from San Francisco's Presidio to Wawona each summer.[1]

Captain Wood finally got a copy of the regulations and some maps through the War Department (from "Wheeler's survey"). His basic charge was to protect the park from vandals, thieves, and trespassers. There were problems with ranchers and homesteaders near or within the park, with tourists and their caterers, and with illicit hunting, but mostly there was a crisis with roiling flocks of sheep ("the curse of these mountains") that left the San Joaquin Valley for the Sierra mountains each spring, and then returned to the San Joaquin in the fall.

Worse, the sheep problem was a fire problem since the herders would set "fire to and burn over the forests" behind them as they drove flocks to winter pasture in the San Joaquin—an ancient practice of transhumance that had emigrated from Mediterranean Europe. The fires meant fresh forage in the spring, and exposing the surface meant an earlier spring melt. Almost from the park's beginning, fire and hoof were seen as the twin threats that challenged administration. Almost from the beginning as well even officers observed that it was nearly impossible to find trees without scorch marks, so endemic was fire, as pervasive as winter snow and summer sun.[2]

The sheep problem was actually a problem with shepherds and could be controlled by eviction. The army eventually devised an ingenious solution by driving the invading flocks out one side of the park, while depositing shepherds on the other side; neither survived well without the other. The project also helped tamp down fire, which to Captain Wood's reckoning was at least as troubling. Echoing progressive thinking of the day, he insisted that "it is impossible" to "affix even an approximate value to the damage wrought by a forest fire, for the consequential damages reach hundreds of years into the future."[3]

Lightning started plenty of fires and ignored orders to cease trespass. But most fires—the ones near valley and grove—were kindled by people, who had explanations as extensive as their flocks. The military dismissed their excuses. The shepherds ("few of whom are American either by birth, citizenship, or sympathy") cared only for their flocks, while the hunters, tramps, and ne'er-do-wells who set or abandoned fires cared only for the moment, and nothing for the future. "The rule which makes a pecuniary penalty for creating forest fires is a mere waste of words, a beating of the thin air. The men who commit this deed are impecunious and can not be touched by a simple fine. They are not the men who observe the laws from a love of order and justice." In August 1892 sheepherders set a fire that burned through chaparral into the park but was beaten down by a "party of students from the University of California."[4]

Yet fire was a publicly visible threat, and granted the embarrassing inability of the commission to handle the 1889 fires, the army needed to show it could meet that challenge. It could fight: it could meet violence with violence. If fire was a threat, it should be fought, and a cavalry unit was seemingly well equipped as a counterforce to match the mobility of fire. Captain Wood thought so, and tallied the number of fires his troops suppressed. But if fire was a military threat, it resembled a sullen insurgency, nurtured by nature, rather than a hostile army that could be defeated in battle.

One strategy was to secure critical sites with a fortified perimeter. In 1892 and 1893 Wood had his troops act on Olmsted's proposal to surround the Mariposa Grove with a road that would serve as a fuelbreak, a project the commission had revived after the recent scare. The road and adjacent tract were cleared of deadwood, which was then piled and burned. Immediately, too, the dedicated captain worked to contain tourists and evict shepherds—keeping the hostiles off the reservation, as it were.[5]

Still, although Interior Department rules forbade fire setting, there was no mechanism by which to fight flames in a methodical way, or as Captain Wood put it, "My men have no facilities here for subduing fires." They had no metal rakes and relied on "gunny sacks" and "brush brooms, etc." They took advantage of open spaces, raking needles and duff into windrows from which they could set a "back burn, and thus check the fire effectually." Fortunately, the fires themselves burned slowly, and unless in a thicket, they could be "approached with impunity." Nor did they generate much convective wind. Nature helped in 1892 with a wet spring and fall that shortened and dampened the fire season. In September 1893 a fire burned into the Tuolumne grove of sequoias and even "touched one of the trees" before it was halted. Another, originating outside the park, was contained by "driving it against the South Fork of the Tuolumne River." The next year there were "no forest fires of any magnitude within the park." A few hand tools, a system of patrol, a transient population either conditioned or compelled not to kindle fires—by such means Yosemite's fire problem would thin into oblivion.[6]

Captain Abram Wood unexpectedly died in April 1894 and was replaced, without much preparation, by Captain G. H. G. Gale, who regarded Wood's "methods as so rational and effective" that he tried to follow them "in every particular." Still, only one member of the command had any experience in the park, which left Captain Gale, the alert captain, to engage in his own reconnaissance and see everything anew with his own eyes. Regarding most of Wood's analyses and actions, he agreed. But not on fire.[7]

The fires he saw burned light, swept away surface needles and windfall, did little harm to the forest overall—virtually every mature tree bore scorch marks, with minor damage from catfaces. "It is a well-known fact that the Indians burned the forest annually." The conditions that had allowed Captain Wood's troops to wrestle fires without even the benefit of rakes was, in truth, the result of incessant burning. The "absolute prevention of fires in these mountains," Gale concluded, "will eventually lead to disastrous results." This was, he admitted, the general view of settlers and especially of the sheepmen, who also "include in their programme the previous destruction of every living thing in the forest within reach of a sheep's teeth." Still, he was able to separate the message from the messenger. Granted California's climate, the endless accumulation

FIGURE 2 Captain Alex Rogers standing beside a fire-scarred tree along the Tioga Road (1897).

of combustibles on the forest floor, and the inevitability of ignition from some source, the fire scene would gallop out of control in short order if all fires were suppressed.[8]

That perception changed when Captain Gale was succeeded by Lt. Colonel S. B. M. Young. Wildfires again loomed large, mostly from abandoned campfires, which Young likened to the "abominable" litter also left. One fire boiled up only two miles from the company's Wawona encampment, to which Col. Young dispatched his "entire available force" but which still required three days to contain. The next summer fires were "unusually numerous" (from the usual causes); one burn threatened the Merced Big Tree Grove. Young also suspected that fires were being set outside the park so that the prevailing westerly winds would drive them across the boundary. A proposal to burn in the spring—kindling dry patches as snows retreated—he dismissed by appealing to the inestimable value of the organic layer ("tree mold") for watershed. Water beat fire. Besides, he had orders. Park policy was to ban burning.[9]

Then the troops left, abruptly, dispatched to the war with Spain. A civilian, J. W. Zevely, seconded from the General Land Office, served as acting superintendent but without the means to enforce any policy, leaving the park "wholly unprotected." Herders returned and "numerous forest fires" with them. In September, the cavalry rode to the rescue with a company of volunteers from Utah. The shepherds used signal fires to warn of approaching patrols. Eventually a large raid captured 24 herders, scattered thousands of sheep, and ended the fires. "There was not another conflagration in the park during the entire season."[10]

Until then, without resources to battle the fires, Acting Superintendent Zevely was left with time to consider what to do about them. Conversations with "old mountaineers," who had lived in the Sierras since gold rush days, convinced him that a fire-exclusion policy was "erroneous." Prior to the park's creation "fires occurred almost every year in all parts of the forest—in fact, they were frequently set by the Indians, but there was so little accumulation on the ground that they were in a great measure harmless, and did not in any sense retard the growth of the forest." There was "not to be found now in the whole forest any tree of great magnitude which has not upon it the marks of fire, yet the trees have in no wise been seriously affected by these burnings." Now, a mere eight years later, needles, leaves, and windfall covered the forest floor to a "depth of from 12 to 18 inches" and large trees had fallen and "the whole mass is highly inflammable." Acting Superintendent Zevely believed it wiser to reverse current policy and begin a program of "systematic burning."[11]

The army issued two brief reports on its short tenure, most of which was spent chasing sheep and fires. Captain E. F. Wilcox, Sixth Cavalry, noted that the fires were "entirely beyond control," yet "did little or no damage," and were ultimately extinguished by fall rainstorms. He recommended "a systematic burning out" along roads. William Forse, 2nd Lt., Third Artillery, thought the fires "the most difficult problem with which I had to contend." They required lots of men and especially tools ("twenty shovels and about one hundred strong iron rakes should be kept in the park at all times"), all under the authority of someone "who understands how to fight fire" and can hire men as needed. He also considered regular burning, which could be done "without great danger by doing so after the first rains or snows." Overall, he considered fighting better than lighting, but without tools the troops were ineffective, and the only way to get tools was to purchase them himself, which made preventive burning more attractive.[12]

So it went until the army left in 1914 and a formal civilian surrogate, the National Park Service (NPS), was established two years later. Official policy was to prevent fires, so the army fought fires. But after a few years on site, or after reading reports of predecessors, cavalry officers tended to favor some program of regular burning throughout the park. The Miwok had burned near annually, newcomers replaced their lost fires, old mountaineers sounded alarms over the threat posed by amassing combustibles. While the gold rush had opened California to the footloose folk of the world, and led to a crash of the native population, it had shifted fire regimes in the Sierra Nevada only slightly. The newcomers had picked up the dropped torch, often citing Indigenous burning as a model. Inveterate burners had multiple reasons to burn, not all based on ecological integrity, and visitors left campfires with abandon. Teasing out the fires from the fire setters, the benevolent effects of burning from those advancing greed and indifference, was tricky.[13]

Major John Bigelow, Jr., laid out the arguments in his 1904 report. "The most difficult and perhaps the most important of my duties in the park is the prevention and extinction of forest fires." Resources, however, did not match policy: he had 36 rakes ("not strong enough to answer the purpose well") and in dry years insufficient troops. But he doubted that facts matched conditions either. His predecessor, Lt. Colonel Garrard, had sided with systematic burning, and so had Garrard's predecessor, Major Hein. The "competent judges" the major had queried were all in favor of burning; only "two persons" seem in any way opposed, for reasons that were not clear to him. The problems lay in practice. The troops departed each fall during the prime time for burning; and having kept fire out for a number of years, it would be a "particularly delicate operation at present or when next attempted." In what appeared a choice between two evils, the major deferred to "experts in forestry." The transfer of the forest reserves to the U.S. Forest Service in 1905 meant those forest experts would be rabid to remove fire.[14]

Meanwhile, some locals took matters into their own hands. On July 8, 1905, fires broke out on both sides of the Yosemite stage and turnpike toll road. Soldiers suppressed them. Two days later the fires returned. They were again extinguished. It seems employees of the stage company were burning about a mile or mile and a half a day, pleading that they were only "back-firing." The soldiers ordered them to stop. Two days later they began again. Captain H. C. Benson, then acting superintendent, determined that in the absence of official burning the locals were determined to do it themselves.[15]

Similar issues appeared on the forest reserves around the park. Writing in 1907 Charles Shinn, forest superintendent of the Sierra National Forest, emphasized "first of all" the importance of "saving every tree, large and small, that we can save from fire, at any cost of time and money." Fire was, "beyond question, the greatest enemy of the forest." When he arrived at the forest, the "Piute system of forestry" based on routine burning was prevalent, and "every one, without exception worth noting, argued openly that fires did no harm at all: but rather did good." Over time, he explained with grim satisfaction, those advocates have recanted.[16]

But along the western front of the Sierra Nevada, well into the 1920s, an insurgency continued to advocate "light burning." It was not just a fight over the control of fires, but a firefight over who should control fire policy and hence public forests. Those on the ground, including officers dispatched from the Presidio, mostly wanted consistent burning. Those setting policies did not. Too many agreed with Bernhard Fernow, an émigré Prussian and Amer-

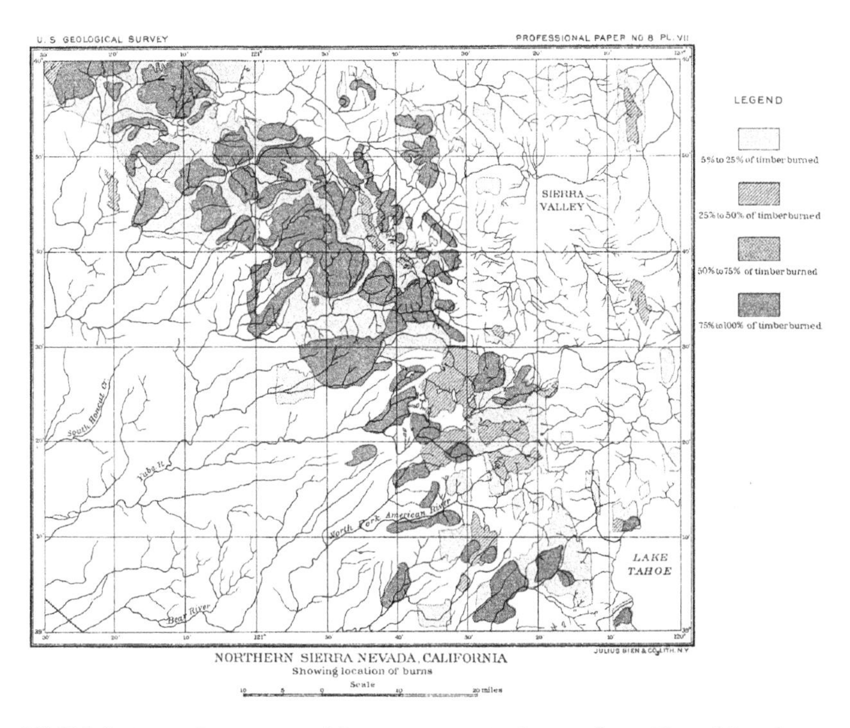

FIGURE 3 Burned area map of forest reserves in the northern Sierra Nevada (1902). The four categories (each a darker shading) are: 5–25% area burned, 25–50%, 50–75%, 75–100%. The cartography confirms what written accounts consistently state: that most of the forest was burned. What the words add that the map elides is that, except in areas subject to heavy mining (and hence slashing), the fires did little damage to standing trees, but encouraged a more varied cover of shrubs and grasses. There is little reason to believe that Yosemite's backcountry burned in significantly different ways.

ica's first professional forester, that the fire scene was largely one of "bad habits and loose morals." Fires were a stigma of the primitive, fires were an index of social disorder. In retrospect, the story of fire policy belongs less with a science of landscape fire (none really existed) than with a Graham Greene novel: good men in a flawed cause, flawed men in a good cause.[17]

Thanks to President Teddy Roosevelt's efforts, the park's boundaries were expanded in 1905 to better approximate natural watersheds. What had been two distinct landscapes, valley and gorge, were absorbed into a single, larger one; what had been forest reserve now reincarnated as national park. The next year California's commission formally dissolved. The army transferred its headquarters from Wawona to Yosemite Valley. The U.S. Cavalry became the effective face of park administration.

Managing a national park was not what the War Department regarded as its primary duty, though using military force to oversee protected sites was common enough throughout Europe's imperium. The 1896 National Academy of Science's Commission on Forests had suggested that the United States look to British India for inspiration and proposed that, on France's example, the country create a forestry corps as part of its military establishment and teach forestry at West Point. Protecting against trespass and fire seemed something well suited to state-sanctioned force. That, however, was not what the War Department wanted; the Spanish-American War and then World War I removed the troops from the park and put them back into the battlefield.

The cavalry certainly proved more effective than California's commission in containing fires. But it further deepened the dissonance between official policy and actual practice. Those in a position to make policy demanded fire's unconditional surrender while those who were in the trenches thought fire made a better ally than an enemy. It was the old distinction between those who declare war and those who have to fight it. Perhaps the cavalry understood the circumstances of Yosemite's firefight only too well.

CHAPTER 3

Gorge and Basin

The road to the Mono Meadow Trailhead that the trekkers will take splits off eastward from the old route that joined Yosemite Valley and Wawona. At Chinquapin, a little less than halfway between valley and grove, an intersection links the formative highway with a paved road to Glacier Point, one of the most celebrated panoramas at Yosemite.

The path to Illilouette begins with that traditional pilgrimage to Glacier Point before splintering off east to the Illilouette. Until then, it travels a well-worn road to a celebrity overlook.

• • •

There are two classic views of Yosemite Valley—one from the bottom and one from the top. In the view from the bottom the approach to the valley suddenly opens, as if through a tunnel, into the long granitic glen of a prelapsarian world. The view from the top looks down onto the granite trough that is the valley, but an equally strong pull is the high panorama that sweeps across the Range of Light. That high perspective is the view from Glacier Point.

There is an oft-reprinted photo that shows President Teddy Roosevelt and John Muir standing, in 1903, on the dome of Glacier Point with Yosemite Falls behind them and the High Sierra crestline limning the distance. The moment is usually interpreted to show

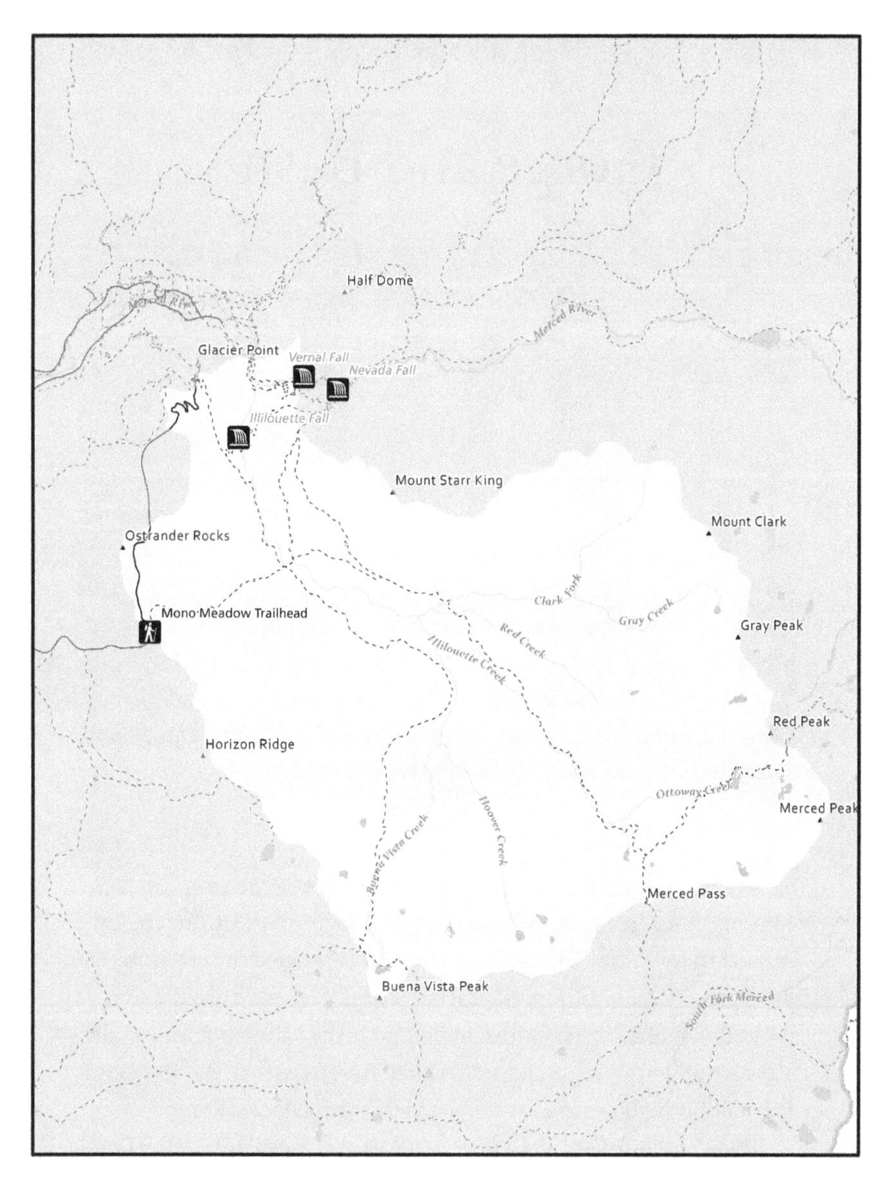

MAP 2 Illilouette Basin.

FIGURE 4 John Muir and Teddy Roosevelt at Glacier Point (1903).

how the prophet of the wilderness, Muir, influenced the advocate of the strenuous life, Roosevelt, to make conservation central to America's political culture. (Two years later the Roosevelt administration oversaw an expansion of the park's boundary to nearly its modern form.) It represents one of the iconic moments of American environmental history—two charismatic men, a landscape of breathtaking monumentality, a moment of revelation.

FIGURE 5 President Theodore Roosevelt with his entourage at the Grizzly Giant, Mariposa Grove (1903).

Inevitably, the story is more convoluted. Muir's was one vision of nature protection; there were others. A second photo of Roosevelt exists with him standing beside that other anchor of Yosemite, the Grizzly Giant in the Mariposa Grove. This time Roosevelt is flanked not by one singular voice, but by a company of aides and luminaries that include the university presidents of Columbia and California, the secretary of the navy, the governor of California, Roosevelt's personal secretary, his personal physician, and two Secret Service agents, all in a photo taken by Joseph LeConte. Roosevelt celebrated not only Muir's perspective but transferred the forest reserves to the Forest Service, used the Antiquities Act to establish national monuments, created America's first wildlife refuge, proclaimed the Grand Canyon the "one great sight every American should see," convened a Country Life Commission chaired by Liberty Bailey and a Governors Conference on Conservation. When Roosevelt toured Yosemite, he did so as a head of state who was also advocating a Square Deal and a New Nationalism. The view from Glacier Point, taken alone, crops out a lot of the narrative of American environmentalism.[1]

The image that places Muir at Glacier Point is valid: its field of vision is a panorama of ice-sculpted landforms. It captures what Muir knew best and did best. The image that removes Muir from Grizzly Giant is equally right, because, for all his passion over the monarchs of the forest, John Muir misunderstood the fire that informed them.

• • •

Glacier Point's very name proclaims the role of ice in molding the park's terrains. Yet, when Muir first visited Yosemite, the concept of an ice age was still new. Geology had only acquired its name in 1783; Louis Agassiz had first announced the idea of continental ice in 1837, and transformed his vision into an argument three years later in *Studies on Glaciers*; James Geikie first consolidated the global evidence into *The Great Ice Age* in 1874, and James Croll followed with an explanation based on oddities in the Earth's orbit with *Climate and Time* in 1875. When John Muir first trekked into the Sierra Nevada in 1868–69, an ice age was still as much a cabinet of curiosities as a coherent concept.

But Muir had grown up in Wisconsin, a landscape repeatedly glaciated and marked by glacial landforms, was a close reader of nature, and recognized the signature of ice on the rocks. "The Sierra Nevada of California may be regarded as one grand wrinkled sheet of glacier records." Josiah Whitney, Boston Brahmin and chief of the California Geological Survey, scorned the notion as nonsense (he thought the valley was the result of faulting) and dismissed Muir as "that shepherd." Muir's interpretation soon prevailed.[2]

Muir joined, then replaced, Galen Clark as Yosemite's prevailing genius loci. He founded the Sierra Club in 1892 to promote its glories. He widely and successfully wrote about nature, and especially Yosemite; his collected works run to 10 volumes. He fought bitterly to preserve Hetch Hetchy, Yosemite's twin valley, from being converted into a reservoir to provide water for San Francisco. John of the Mountains became a national prophet for wilderness.

In *The Mountains of California,* assembled from assorted essays in 1894, he traces virtually every landscape feature to the action of the ice. The trough valleys scoured by glaciers, the rounded domes smoothed by ice sheets, the serrated crestline etched by cirques—the ice linked not only geology but biology, with the distribution of forests and sequoias determined by the soils and moraines cached, deposited, or missed by ice, and it is his vision of the ice that holds the separate essays together in the book. The residual glaciers in the High Sierra loom above everything, a token of the transcendent presence of the ice. The valley is a museum and laboratory of glacial ice. The grove exists, as do other sequoia enclaves, because the last surge of ice had bypassed their locales.

Muir recorded fire with the same close attention with which he read nature everywhere. While a young boy, he exulted in the "magnificent brush fires" used in clearing land for the family's Wisconsin farm, though his father compared tossing branches into the flames as souls tossed into hellfire. He saw how fire created the "oak openings" that abounded with prairie grass and forbs; a few trees survived the annual flame-sweeping because they had "the good fortune to grow on a bare spot at the door of a fox or badger den, or between straggling grass-tufts wide apart on the poorest sandy soil." Without fire the prairies would have been "covered by the heaviest forests." Even a few years' lack of fire was enough to allow oaks and other trees to overrun them.[3]

In the Sierra Nevada he saw evidence of fire nearly everywhere and marveled at the adaptations displayed. Chaparral fires took the place of prairie fires in sweeping trees away. He noted how "Indians walk softly" on the land, save for the "fires they made to improve their hunting grounds." He called attention to the "admirable adaptation" of the pinyon pine to the "fire-swept regions where alone it is found." He tallied the many ways the limber pine was subject to fire and adapted to it. The "inviting openness of the Sierra Woods," he wrote, "is one of their most distinguishing characteristics"; trees stood apart or in clumps, leaving a "park-like surface"—the outcome of incessant flames sweeping the forest floor.[4]

In a dramatic passage he described a fire "racing up the steep chaparral-covered slopes of the East Fork canyon with passionate enthusiasm in a broad cataract of flames" until it reached the "deep forest," where the "ungovernable flood became calm like a torrent entering a lake, creeping and spreading beneath the trees." There was "no danger of being chased and hemmed in, for in the main forest belt of the Sierra, even when swift winds are blowing, fires seldom or never sweep over the trees in broad all-embracing sheets" as they did in the Rockies and Cascades. "Here they creep from tree to tree with tranquil deliberation, allowing close observation."[5]

Yet where his observations on ice led him heavenward, his experiences with fire took him down to a hellish place where fire joined other misdeeds of humans toward Creation. Muir reckoned that "notwithstanding the immense quantities of timber cut every year," from "five to ten times as much is destroyed as is used," chiefly by fires. Those fires accompanied prospectors, loggers, and especially shepherds in marring the wonders of nature. What escaped the maws of sheep succumbed to the flames of shepherds, "which are set everywhere during the dry autumn" and "are far more universal and destructive than would be guessed." Those "destructive" fires ramble "through nearly the entire forest belt of the range, from one extremity to the other, consuming not only the underbrush, but the young trees and seedlings on which the permanence of the forest depends; thus setting in motion a long train of evils." What Muir saw as future forest, however, a future generation would see as unwanted fuels, as ugly as an oil spill, ready to power megafires.[6]

If plants showed intricate adaptations to fire, those accommodations were—to Muir's mind and to most of the educated elite of the time—prompted out of self-defense. The trees would thrive better if those insults were removed, much as humans would if malaria were abolished. There was no sense that the greatest threat posed might not be fire's presence but its absence. Removing fire from the valley had allowed it to overflow with trees, just as happened with Wisconsin prairies. Removing fire from the grove encouraged white fir, incense cedar, and windfall to fill the clearings between the Big Trees and replace benign fire with malignant.

The Big Trees were where Muir's thoughts on fire best converged. The *Sequoia gigantea* was the "monarch of monarchs." Fire, he concluded, was "the arch destroyer of our forests, and sequoia forests suffer most of all. The young trees are most easily fire killed; the old are most easily burned, and the prostrate trunks, which *never rot* . . . are reduced to ashes." In 1875 he "examined every sequoia grove in the range [excepting the Calaveras], together with the main belt extending across the basins of Kaweah and Tule, and found everywhere the most deplorable waste from this cause [fire]."[7]

While he was at the Kaweah River, the forest was burning. "And as fire, whether occurring naturally by lightning or through the agency of man, is the great master-scourge of forests, and especially of sequoias," he seized the opportunity "to study the methods of its destruction." The fires burned up the young saplings and hollowed the fallen giants. But within the grove, fire could "only creep" because there is "no generally distributed fire food . . . on which they can move rapidly." It could not have escaped him that there was no fuel because the fires were so frequent, but he reasoned the groves would still be healthier without them. Elsewhere, he observed 86 sequoia saplings "upon a piece of fresh ground prepared for their reception by fire," but he concluded there were other, better options by which to access bare soil that did not involve burning. Even when fire helped it harmed.[8]

Muir was not alone. When visiting the Kaweah grove in 1891, Gifford Pinchot admired how scorch mixed with red bark, an effect "brilliant beyond words," then dismissed the "highly decorative but equally undesirable fires," which bulked so large in the minds of settlers, most of them "Eastern tenderfeet." One informed Pinchot that the newcomers had saved the Big Trees from burning up 29 times in the last five years. "Which might naturally have raised the question," Pinchot mused, "Who saved them during the remaining three or four thousand years of their age?" Pinchot might mock the newcomers, but the would-be westerner had no better understanding of fire's ecology in the grove. No more than Muir was he prepared to pick up the other end of the stick. Fire was vivid, ready to animate a larger message of conservation. Its nuances mattered less than its power to rally political enthusiasm.[9]

John Muir's passion for Yosemite needs no defense. The transition from native to newcomer was harsher than it should have been, but would never be easy, and Muir more than anyone else defended valley and grove against the bottomless depredations the newcomers wrought. (His disparaging remarks about natives focused on their broken, degraded, and—always a telling trait for Muir—unwashed demeanor, which echoed his disgust toward slovenly shepherds. "Perhaps if I knew them better I should like them better.") He viewed fire through the prism of an era of profound *un*settlement, as a further act of senseless vandalism and greed. That forests could absorb fire demonstrated the deep durability of nature. But they would be better off without it, as people would without war or smallpox.[10]

• • •

What is striking is how often people on the ground favored fire while intellectuals, reformers, and officials condemned it. Fire reeked of primitivism; a rational society would find alternatives to it and replace mindless tradition and superstition with enlightened science. Yet there was no genuine science of landscape fire, and would not be for another 50–60 years after Muir's death. There were only opinions voiced by the educated and powerful which reflected shared values about how they understood the world worked and which their holders assumed must represent positive knowledge.

Besides, flames and landscapes blasted into ash were too useful to help animate other messages. While ordinary citizens seemed

FIGURE 6 Charles Walcott, director of the U.S. Geological Survey (USGS), in 1897 standing on the trunk of a large tree toppled after fire ate a cavity through its base. This kind of biotic abrasion by seemingly benign surface fires was another charge critics leveled against light burning. From 1898 to 1902 the USGS mapped existing and proposed forest reserves and included burned area in those maps. A century later the USGS absorbed the National Biological Survey and returned to Yosemite to research fire as well as geology.

perfectly capable of distinguishing between a river in flood and an irrigated field, publicists and promoters apparently believed that distinguishing between a wild and a controlled fire was beyond them, that any complexity weakened the power of fire to mobilize public enthusiasms, that it was unwise to separate message from messenger. The pyrophobes were lumpers, unwilling to parse fire into its many varieties. One outcome was that fire was too often exploited to promote programs that actually lessened the ability to live with fire.

What happened to the story behind the photo of Roosevelt and Muir at Glacier Point happened also with American environmentalism and with fire. What didn't fit preconceptions was cropped out. The narrative simplified into Manichaean conflicts between right and wrong, good or evil, instead of complicated choices and trade-offs involving good fire and bad. The light burning controversy assumed the same character as the controversy over O'Shaughnessy Dam at Hetch Hetchy. There could only be one right choice, made of whole cloth, instead of patchwork quilts of practices adapted to the peculiarities of particular places. John Muir's views of wilderness helped create the park. His view of fire complicated its management.

Instead of a panorama of fires burning freely across a fire-accommodating backcountry, of fires renewing the valley's meadows, and of fires cleansing the sequoia-temple groves, the only fire visible from Glacier Point was the ember cascade that, nightly, for the entertainment of its customers, the Curry Company poured over the rim as the Firefall.

• • •

Glacier Point exists to look down along the trench of the valley and to the source glacial fields beyond. At 7,214 feet, the point is high, the view deep. Granite walls channel the eye as they do water; the Sierra crest inscribes the farthest field of vision.

Yet pivot right, to the south, and the perspective changes. Glacier Point is the terminus to Illilouette Ridge; the forested watershed of the Illilouette Basin fills the horizon between the road and the valley rim. The valley, punctuated by El Capitan, the Three Brothers, Half Dome, and Yosemite Falls, speaks to a geological epoch of ice. The basin, with no comparable monuments to organize its field of vision, testifies to an Anthropocene powered by combustion. Once, ice filled the valley. Now, smoke does.

CHAPTER 4

Road and Trail

The Mono Meadow trailhead lies about five miles south of Glacier Point. It spalls off from the main tourist route and heads eastward from Horizon Ridge into Yosemite's wilderness and the drainage basin of Illilouette Creek. Some 5,000 visitors drive to Glacier Point on a summer's day; backcountry parties to places like the Illilouette are limited to no more than 15 persons.

Road and trail are two paths that diverged in Yosemite's woods. But equally to the point, they divide history. Take one, and it is impossible to double back in time and choose the other. If you want to change direction, you have to go from where you are. History isn't like unscrewing a light bulb you previously screwed in. You can't rewind history like a tape. The decision paths Yosemite scratched early into its surface eroded over time into ruts and washboarded slopes.

The determination to abolish fire was widened from a footpath to a road that was eventually paved and became the thoroughfare for the park's future choices not only about what Yosemite meant, but about how fire fit into that vision. Once started, fire suppression became more difficult since fuels blossomed unchecked, but it also became more difficult to retrace the steps that led to that outcome and find an alternative, which made the existing road the path of least resistance. Fire control led to further fire control. Awkward mismatches between what people saw and what they were told to do were forgotten as newly forged cultures, funding, the examples of neighbors, and momentum discouraged other options. It would be impossible to take the park into a different direction from Glacier Point itself. An alternative future required a path that could break away from the paved past before it reached that end.

Like the trek, the trail into the Illilouette offers a chance to step aside from the inherited flow of history. One by one the trekkers gather at the trailhead. Packers cinch their duffels and collective camp gear onto mules. The storm that has kindled three fires in Yosemite has started two in Sequoia-Kings Canyon National Park, its Sierra sibling. Yosemite extinguishes its fires, while the two at Sequoia-Kings evade initial attack. Sun shines through the fir and pine, and a filmy veil of smoke drifts from the south.

• • •

The army era left its imprint. Its civilian replacement, the National Park Service established in 1916, kept the cavalry's campaign hat, gray shirt and epaulettes, pantaloons, and policies, along with something of its divided mind over whether it should fight fires or light them. In practice, there were scant resources to do either; most of the new agency's energies went toward promoting public enthusiasm for the parks, which translated into wider access and more spacious tourist facilities.

The infrastructure the NPS inherited was ill equipped to fight more than the most elemental fires. Even the troopers had struggled to find enough hand tools. In 1904 the Interior Department furnished them with 24 pitch rakes, 24 rake handles, 12 shovels (long handled), 12 large hoes, 12 axes (double bitted), 6 hatchets, and 2 gross of friction matches. Otherwise they used pine boughs and blankets for swatting flames. Park rangers were no better equipped, and had fewer people to call upon; for large fires Yosemite had to look outside agency ranks. As the army had assisted the Forest Service along their shared border, so the Forest Service assisted the Park Service.[1]

Given its lean tool caches, the park's key strategy was not to suppress fires but to prevent them. The principal fire starters were expelled; careless tourists were congregated into campgrounds; paved routes reduced the prevalence of roadside fires from matches and cigarettes tossed by sightseers. In the early years, while the forests and prairies were still largely shaped by a long history of regular firing, fires could be subdued by direct action or burnouts along roads or rivers—not pleasant work, often lasting for several days, but still doable with the fire manpower at the park's disposal. Unless controlled burning was revived, however, those firesheds steadily filled with combustibles that would burn from wildfire and would worsen as the decades clocked by.

Compared to the Forest Service, the Park Service was less a national institution than a gaggle of baronies ruled by semiautonomous superintendents for whom local circumstances counted for more than national directives. Mostly, this meant Yosemite had to engage with the Forest Service, but even as the two agencies converged on a policy of fire suppression, they behaved like uneven suns in a binary orbit. The national forests held a much larger estate, boasted a budget for fire, and had academic forestry to promulgate justifications for fire's removal; as often as not, firefighting became the agency's public face. Even before the trauma of the 1910 fires, the Forest Service made fire control fundamental to its mission. In his 1899 *Primer of Forestry* Gifford Pinchot thundered that "of all the foes that attack the woodlands of North America, no other is so terrible as fire." After he became chief forester of the Forest Service, the *Use Book*, the manual he authored to guide administration of the national forests, listed fire control as the first of three duties of its ranger corps.[2]

The Park Service, still sloughing off its cavalry tutelage, lacked comparable size, political punch, and a conviction that fire control had to be a foundation of its administration.

• • •

Yosemite's conflicting opinions regarding fire were part of a global discourse set in motion by colonization. What happened in Yosemite with the park commission and the cavalry era subsequently propagated throughout California under the rubric of "light burning," appeared throughout America as a debate between traditional fire use and a modernizing determination to eliminate fire, and reincarnated throughout the European imperium from Algeria to India, and Cape Colony to Australia. The spectacle of routine landscape burning by natives and newcomers was nearly everywhere. British foresters in India plunged into a virtually identical controversy over fire policy in tandem with the Yosemite commissioners.

But Yosemite did more than publicize Indigenous burning—accounts were already innumerable. A few European explorers like Giovanni da Verrazano had seen the New World through its smoke before they spied its land, and the Spanish discovery of California reported smoke before it traced the contours of the coast. (Pedro Cabrillo named San Diego the Bay of Smokes.) What made Yosemite different was that establishing it as a park made the choice about appropriate fire practices a matter of policy and public debate. Just as Yosemite inaugurated a conversation about how a nature preserve should be managed, so it kindled a quasi-formal discourse about how fire in reserves should be handled, what may well be the first such recorded discussion anywhere.

In 1890, the year Interior Secretary Noble promulgated a fire ban on public lands, John Wesley Powell, then director of the U.S. Geological Survey, tried to convince him that the Indigenous burning Powell had witnessed in Utah, and that his crews had mapped, offered a model for forest protection. In his initial survey (published in 1878), Powell had envisioned the burning as a threat to forests, and hence to watersheds, and so to the irrigation agriculture that he believed was the basis for American settlement in the arid West. But further study led him to reverse himself and he presented his revised case to Noble. Instead, the secretary followed the advice of the Bureau of Forestry, whose chief, Bernhard Fernow, scorned Powell's appeals as unscientific vandalism. ("Major Powell launched into a long dissertation to show that . . . the best thing to do for the Rocky Mountain forests was to burn then down.") That Powell was then also director of the Bureau of American Ethnology and an ardent student of the Paiutes is likely the origin of the expression "Paiute forestry" as a term of derision. But Yosemite's fire debate was already well underway and soon expanded across the many expressions of light burning that spanned the Sierra Nevada and beyond.[3]

Light burners wanted to continue folk fire practices, as adapted from the natives. They were an ungainly lot—William Hall, state engineer (and member of the Yosemite Park commission); George Hoxie, writing for *Sunset* magazine; timber owners like T. B. White and the Southern Pacific Railroad; the novelist Stewart Edward White; the poet Joaquin Miller, who claimed authority from having lived for a while among the Modocs. Beyond their shared cause for routine burning, they had little in common.[4]

Against them stood Progressive reformers and a newly professionalized forestry that looked to Europe and Europe's imperium for inspiration. They were as uniform as light burners were diverse. They claimed authority; they spoke with one voice and tolerated little dissent; they saw fire as a proxy for the entirety of state-sponsored conservation. When Secretary of the Interior Richard Ballinger went on the record as favoring light burning—the same Ballinger whose quarrel with Gifford Pinchot had led to Pinchot's firing—the issue got polarized and politicized. On Pinchot's example, those who argued for fire control insisted that people choose one side or the other. There could be no middle ground.

The brouhaha over light burning was not peculiar to the American West: in one flavor or another, it was present throughout the country, most famously in the Southeast (under the guise of "woods burning"). Nor was it special to the United States. It flared up wherever European expansion pushed into new lands, loosing axe and fire, and so sparked efforts to contain those outbreaks. It was convenient to confuse fires that boiled over from feckless logging and land clearing with those customary fires that had created the valued forests in the first place. Fire symbolized disorder, the irrational, the precivilized. Fire control replied with the might of Enlightenment science and technology.

This blindness toward fire's value was typical among European intellectuals, and they bestowed the same disdain toward routine fire use onto Europe's fire-besotted peasantry as they did on tribes in far-flung colonies. European elites treated the traditional fire practices of Finnish farmers, Greek pastoralists, and German peasants in the Black Forest with the same mix of outrage and dismay they did tribes in India and Australia. American foresters, at one point, even enlisted the advice of anthropologists and employed a psychologist to understand why ruralites, especially in the Southeast, uneducated people mired in poverty and a repugnant racial caste system, insisted that fire improved lands when the nominal science of the day condemned it.[5]

Yet with regard to fire, central Europe was anomalous, not normative, and a silviculture founded on temperate landscapes found itself outraged as it discovered that, overall, the world was inclined to burn, and that people everywhere were happy to help the burning along. When propagated under colonial conditions these perceptions worsened because unruly fires were taken as a metric of social messiness and sullen hostility: controlling fire was a means of controlling restive peoples. The French in Corsica and north Africa, the Dutch in Indonesia and the Cape of Good Hope, the British in India, Africa, Canada, and Australia all experienced comparable controversies, and everywhere the discourse cleaved along the same issue: Should forest protection be based on lighting fires or on fighting them?

The first formal forestry conference, staged in British India in 1878, opened with the question, Is fire control possible? And if possible, is it desirable? The audience split in its judgment. Authorities, academics, ministries of agriculture, and officials charged with public order argued that fire control was foundational, and however onerous had to be upheld. Those on the ground—settlers, Indigenes, low-echelon agents of state institutions—argued that fire exclusion was impossible and dangerous. While settlement shocked Indigenous societies, the newcomers mostly adapted their fire practices such that similar fire regimes persisted. A boundary survey of the Plumas forest reserve in 1904 observed that "the white man"—picking up the torch dropped by the natives, who were "accustomed to burning the forest"—has "come to think the fire as a part of the forest, and a beneficial part at that. All classes share in this view, and all set fires, sheepmen and cattlemen on the open range, miners,

FIGURE 7 Light burning in action on the Plumas National Forest, October 1910.

FIGURE 8 Systematic fire protection demonstrates its firepower. Originally, the emphasis was on rapid detection and attack through lookouts, telephone lines, roads, and trails. But as the photo shows, the force needed to hold flame increasingly came through machines.

lumbermen, ranchmen, sportsmen, and campers." The fires, it was noted, "seem to do little damage."[6]

That was true throughout the Sierra Nevada. Two prominent Forest Service officials, S. B. Show and E. I. Kotok, noted that when the forest reserves were created, forest burning was "an established practice," and that locals considered any attempt at fire's exclusion "preposterous" and doomed to backfire. The two foresters openly admitted that fire control was possible because "the forests were still open as a result of the repeated fires of the past." Only after fire's removal, as young undergrowth flourished and "the amount of inflammable material in the forests increased greatly," did agitation to renew widespread burning go public. The burners' threat was double: "not only because of their direct action, but even more so because of their open preaching of fire."[7]

What Show and Kotok saw as a future forest needed to replace old trees and ensure "forest influences," a generation 40 years later saw as an overgrown tangle, a fuel cluster bomb ready to explode.

Same scene, different visions; same evidence, contrasting meanings. Show and Kotok assumed that research, technology, and political will would assure that fire control could continue to ramp up faster than the unfavorable fire conditions that they regarded as necessary to assure a productive forest. (Similar arguments are being made today that, to address the climate crisis, we must plant millions of trees to soak carbon dioxide out of the atmosphere, while assuming fire control can hold the line against the increasing risk of wildfire they pose.) By the 1960s that expectation was becoming hard to believe, and by the 2010s feral megafires overpowered any counterforce humans could muster and had become a symbol of a fast-marching Apocalypse with California as its flaming front. Early twentieth-century forestry had forced a gestalt-like switch in perception. Early twenty-first-century environmentalism forced another.

Fire as fire was rarely the point: it was what fire enabled and what it represented. Ultimately, the controversy over suitable fire regimes may be best understood as a contest between tradition and modernity. Along with railroads, telegraph lines, and modern science, a new order of fire protection was seen as a rationalizing of folk ways, premodern commerce, and tribal politics. Traditionalists pointed to customary fire practices that had the practical experience of centuries. Modernizers denounced that tradition as both a means and an expression of backwardness that retarded progress. They regarded fire's mere presence as a stigma of primitivism and its use as an invitation to anarchy. Each side dismissed the other as misguided if not malicious. Reformers exploited the visceral power of fire to animate their larger message, much as activists use it today to promote the urgency of the climate crisis.

• • •

Not everyone advocating for light burning was a paragon of clear reasoning, and not all those arguing against it were racist idiots. Consider Coert duBois who, because of a change in family fortunes, was unable to attend Yale as his father and older brother had and who instead trained at the Biltmore Forestry School before joining the Bureau of Forestry in 1900. A very "young, impressionable, and enthusiastic idealist," inspired by the conservation "crusade," he bolted into the fray "head down and tail up." In 1910 he had to interrupt his honeymoon to fight the Stony Creek fire at Lake Tahoe. The next year, at the age of 29, he became the regional forester for California and drafted a fire control plan for the Stanislaus National Forest (which borders Yosemite National Park on its northern half). Three years later he "retired to [his] cell, like John Bunyan, to write THE book on Fire Protection," what became *Systematic Fire Protection in the California Forests*. Light burners had nothing like it. In 1917 he served as a major with the Tenth Engineers in France, then returned to the Forest Service after the armistice, and arranged for the first experimental use of surplus military biplanes for aerial fire reconnaissance. He resigned from the agency in 1919 to transfer to the Consular Service, where he remained until 1948.[8]

Prior to his retirement dinner, a reporter with the *Sausalito News* interviewed him "as the leader and organizer of the most comprehensive fire-prevention service in existence in the west, and particularly as the principal opponent of the so-called 'light-burning' theory of forest protection." The future of California, duBois explained, depended "first and foremost upon the protection of the young growth in our forests from fire." Yet before the agency arrived the state's innumerable fires "were viewed with amazing unconcern." The reason: "the pernicious, ill-advised and destructive light burning theory." If it had continued, in 50 years' time California would have "no forest at all." Where light burners saw a forest that existed because of routine burning, he saw one that existed in spite of it. *Systematic Fire Protection in the California Forests*—"the hardest mental work I ever did, and . . . the most important contribution to the public I ever made"—was his response to the crisis of California's forests. It became the basis for fire planning throughout the Forest Service, and along with its successors,

America's primary contribution to global forestry. Two years after he departed, the Forest Service held its first conference on fire at Mather Field outside Sacramento. What happened in California didn't stay in California.[9]

The Forest Service rallied against light burning not just for what it meant in the Sierra Nevada, but for how it defined fire protection and state-sponsored forestry everywhere. Pinchot's successor as chief, Henry Graves, helped nationalize the cause by explicitly condemning light burning, and his successor, William Greeley, who had led the fight against the Big Blowup, disdained the proposal with the long-standing charge that it was mere "Piute forestry," unbecoming a rising global power. The controversy simmered and occasionally flared for over a decade until, following a state-commissioned board of inquiry, light burning was officially euthanized in 1923. Even Aldo Leopold, then engaged in establishing a fire protection system for the national forests of the Southwest, argued against the insurgency. "The Forest Service policy of absolutely preventing forest fires insofar as humanly possible is directly threatened by the light-burning propaganda."[10]

In his memoir, *Trail Blazers*, Coert duBois focused only on those early, animated years in the fight for conservation. Fifty years after he left the agency, conservation was morphing into preservation, what he had regarded as tomorrow's timber looked to another generation like today's tinder, and the bold trail he and his undaunted cohort had blazed seemed to have taken a wrong fork.

• • •

The Park Service was less monolithic, politically more feudal and less able to impose a common policy on its refractory barons. In Sequoia National Park Col. John White advocated for light burning, fashioned a program that relied on controlled burning in multiple settings, and was willing to go head-to-head with Assistant Director Horace Albright, who was hostile toward any strategy other than full-throttle suppression and notorious for his vindictiveness toward critics. California's Board of Forestry might agree and condemn light burning as anathema, but California's fires replied in their own way. The 1924 season marks, in many ways, the onset of a century of big burns, campaign fires, fire sieges, and megafires. Even pulp writer Zane Grey published a "warning to California" that uncontrolled wildfires threatened its way of life. Superintendent White cleared sequoia groves and practiced selective use of "light burning" in the frontcountry and "let burning" in the backcountry.[11]

Yosemite did nothing comparable, and neither did any other Park Service authority. Where remote fires were left to burn, it was due to lack of men, equipment, and access to put them out. Instead, Yosemite felt the 1924 season through a wave train of nasty fires that threatened its western border, especially along the Merced River. The worst started "apparently by a convict crew" constructing a road to El Portal. Most of the outbreaks were the primary responsibility of the Forest Service and Yosemite Lumber Company; each night a new line would be scratched out, and each day winds would hurl flame across it. The park became actively involved when the Forest Service attended to a fire rushing along the north side of the Merced River, while the park looked to the flames on the south side. The park rallied its employees, along with men from the Curry Camping Company and the Yosemite National Park Company, and dispatched them to Chinquapin, where they were to backfire along the railroad, well in advance of the main fire. It was open country because most of the landscape had been "logged over," though the railway complicated matters because ties and trestles were "saturated with oil." After three days a relief crew was organized. But the line eventually held. While nothing in the park proper burned, the threat had been real.[12]

The Forest Service and Park Service had cooperated well. They agreed to split the costs of suppression. But the national forest had to contend with logging, both the slash left behind and the timber berths still available for harvest, and had adapted its strategy accordingly. The park, as always, attended to tourists. The smoke from the fires had flooded the valley, and "visitors at the camps and hotels, as well as those in the public camp grounds, left the park in great numbers, thinking that they would be in danger in remaining or would be conscripted for fire fighting" (none were). Thanks to "erroneous publicity given by the newspapers," travel to the park plummeted. The season ended with a two-acre fire near the park boundary at the Mariposa Grove that the park and national forest extinguished without any damage.[13]

Yosemite was ill equipped for such threats. The Park Service lacked funds as well as a national model or guidance for firefighting (on the Merced River fires, the Forest Service had sent an adviser). Not until 1922 did Congress even grant the Park Service a fire-specific budget, though without the kind of emergency funding available to the Forest Service. Whatever it spent over its meager allotment would have to come out of its regular appropriation. The 1926 fires in Glacier National Park were so onerous that the agency had to close some parks altogether to pay the costs of suppression (it had a fire budget of $38,000, but a suppression bill of $230,000). "The Glacier fires in 1926," Hal Rothman determined, "were a disaster for the National Park Service." It found itself not only unable to halt the fires but subject to blistering political attacks. A big fire in a single park had rippled through the entire system in unhappy ways.

Two years later the agency found itself at another fork in fire's backcountry trails. The path it chose in 1928 determined its fire story for the next 40 years.[14]

• • •

The immediate provocation involved a fire that rambled into Sequoia National Park from the surrounding national forest and state lands. Under White's direction, the park had fought it with the capabilities it had, but White's persistent and outspoken interest in light burning allowed critics to attack the park as soft on fire control. The NPS needed a public display of enthusiasm for aggressive fire protection, whether or not it had men and matériel to conduct it.

All this happened while national developments further pressured the Park Service to come into alignment with its often rival, sometimes colleague the Forest Service. Congress passed the McSweeney-McNary Act, which consolidated all federal efforts in fire research under the Forest Service, while, as part of its obsession with efficiency, the Hoover administration created a Forest Protection Board to unite all the federal agencies with some responsibility for forests in ways that would eliminate "duplication." Inevitably, the Forest Service chaired and dominated the project; and unsurprisingly, the board considered fire its prime problem. The National Park Service was expected to participate and then comply.[15]

It responded to both national and California pressures by hiring John D. Coffman from the Forest Service to organize a national fire program. Coffman had cut his teeth in the incendiary-ridden Mendocino National Forest, had fought through the light burning controversy on part of its home turf, and brought to the Park Service forestry's passion for fire control with the Forest Service's capacity for organization. Before the year ended he submitted, as required by the board, a Forest Protection Requirements report, and he addressed the still festering crisis at Glacier by applying Forest Service methods, though adapted to the Park Service mission, to craft a model fire plan. It was immediately tested by another blowup at Glacier, the Half Moon fire, which not only managed to frustrate all efforts at containment but leaped the park boundary, and once more seemed to bring the National Park Service's competence into disrepute.[16]

Meanwhile, Yosemite asserted its claim for attention with a large fire around the Merced Grove in August for which it had to muster a 200-man fire camp. A fire plan followed. Since fire was viewed as something that happened during the summer, in some years and not in others, it was massaged into other duties of the ranger corps, rallying additional help from staff or the Forest Service or local sources as needed. In 1930 the park began to record fires and map them on a master atlas. In 1931 it got a lookout at Crane Flat. It obtained a fire truck for building fires. Its roads, especially around the Mariposa

FIGURE 9 John Coffman (front right), at 1931 staff meeting at Sequoia and General Grant National Parks.

Grove, doubled as fuelbreaks. The park did no preparation and mitigation work other than sponsoring an annual fire training exercise. It had a dedicated fire staff of exactly no one.

Then, in 1933, as a centerpiece of freshly elected president Franklin Roosevelt's administration, the Civilian Conservation Corps (CCC) came to Yosemite. The CCC was a cornerstone of the New Deal, and NPS Director Horace Albright ensured that the camps became the means for erecting an infrastructure that would otherwise have taken the ever-starved agency decades to install. That Albright assigned John Coffman as lead between the CCC and the Park Service, between "emergency forestry" (i.e., fire) and the general "emergency conservation work" that was the informing theme of the CCC, guaranteed that fire protection would claim a healthy tithe. Yosemite had five camps in all, each of a hundred enrollees.[17]

FIGURE 10 Alder Creek fire, 1931, the year of the park's first lookout and the beginning of fire records.

It was heady stuff for an agency that often struggled to find enough shovels and rakes for the fire crews it assembled from time to time. Still, the Forest Service oversaw roughly half of all the camps established nationally until the CCC disbanded in 1942, as the country left the Great Depression for World War II. It is estimated that perhaps half of all the work the camps did related to fire—all the roads and trails, all the lookouts, all the telephone wire strung through the backcountry, all the roadside cleanups, all the fire caches and remote toolsheds, not to mention all the bodies available for fire call. Within the national park system, across the roughly eight years it flourished, the CCC contributed an estimated 837,783 workdays of fire-related projects apart from actual fire suppression and 688,255 workdays of dedicated firefighting.[18]

California was notable for the expansiveness of its vision, of which the Ponderosa Way is a marvelous emblem. Here the Forest Service embarked on an 750-mile-long fuelbreak and truck trail

FIGURE 11 CCC fire training school, Wawona (1941).

that extended the entire length of the Sierra Nevada from Bakersfield to Redding and was intended to permanently segregate the "brush" from the "timber." Such a project was unthinkable without the bottomless labor of the CCC, and perhaps unimaginable any place outside California with its long, dry summers, its fire-thirsty biotas, and explosive autumn winds. So powerful, and so suddenly, did the CCC impress itself on federal land agencies that in 1935 Chief Forester Gus Silcox promulgated what became known as the 10 a.m. policy. "An experiment on a continental scale," the proclamation established a single national standard: the control of every fire by 10 a.m. the morning following its report. The conception was utterly implausible without the CCC to underwrite it.[19]

The 10 a.m. policy laid down the goals of national fire strategizing. Subsequent fire plans at Yosemite sought to meet it. The CCC erected another lookout at Henness Ridge and a fire guard station at Miguel Meadow. The bulked-up fire program needed better guidance at those parks most prone to fires, which led, on Coffman's example, to the select hiring of foresters. All fire control programs

FIGURE 12 Fire road built by CCC (1934).

were relocated to the responsibility of each park's chief ranger, where fire-related chores belonged. Still, fire was something done by people mostly hired for other reasons and listed under "other duties as assigned." Yosemite recruited Emil Ernst, who paradoxically proceeded to document the changes wrought by fire exclusion.

So sudden and overwhelming could the transformation be that critics arose to protest the apparent subordination of all other values to the imperatives of fire control. It was possible to be too aggressive in cleaning up forest floors, punching fire roads into backcountry, and erecting lookouts, and CCC legions could alarm those who treasured wilderness and wildlife especially. Coffman drafted guidelines that adapted general principles, best exemplified by the Forest Service, to the special mission of the Park Service. But there was no compromise with the need for fire control. Coffman argued against both light burning ("a practice [that] cannot be tolerated in the national parks") and let burning ("there are extremely few areas where any fire starting is not a threat to high values"). When the ever-vexing issue of encroaching forest overgrowing the valley (and its vistas) revived, Coffman dismissed Ahwahneechee precedent as a "haphazard practice" and "strenuously oppose[d] any consideration of broadcast burning within the forest growth of Yosemite Valley." An alternative now existed. With the CCC and a growing reliance on engines, the parks could hope to match the power of flames with muscle and machines.[20]

• • •

The CCC years produced a bureaucratic sugar high, and the austerity of the war years and postwar recession that followed made the resulting crash seem even harsher. The Park Service lost 25 percent of its staff. After peace arrived, visitation, impounded by wartime rationing, broke through the dam and began to flood the park; and as always the park attended first to its primary constituency. But fire protection also found new purposes and resources to continue its project.

FIGURE 13 Fire truck and crew (1941).

The mechanical might provided by war-surplus machinery replaced the massed muscle of the CCC. The Forest Service had priority access to aircraft, bulldozers, transport trucks, jeeps—what couldn't be applied directly to firefighting could be modified by the addition of plows and pumps. The Park Service acquired some of the surplus, but accessed most of what it needed through cooperative agreements with the Forest Service, and once such machines were embedded into the infrastructure of fire control, civilian alternatives could supplement and replace them. The use of war machines against fire reinforced the notion of fire as an enemy to be fought (the other "Red Menace"), as America entered a Cold War on fire. Still, at Yosemite, through the next three decades, fires were a minor nuisance and easily contained, with a few exceptions like the 1948 Rancheria fire.

Rancheria showed the fusion of difficulty and luck that has characterized Yosemite's backcountry fire program. A Board of Review, chaired by Coffman, identified the factors that made the fire hard to find, awkward to control, and expensive to extinguish. The fire—actually, two—burned from September 9 to 21. Terrain and fuels resembled that around Arch Rock. The fires were tricky to locate: their smokes were invisible to existing park and surrounding national forest lookouts, and were spotted by a Park Service trail crew foreman, who with another man and two shovels had made initial attack. Four fishermen helped, contributing a "small trenching shovel" and otherwise worked through the night, using their hands and feet. One fire was contained, the other broke free. Eventually the Crane Flat lookout reported smoke on Rancheria Mountain, and the buildup began.[21]

The September starts left the park shorthanded of seasonal personnel. Overall, it contributed 18 rangers, 3 naturalists, 3 fire control aides, 43 blister rust workers, and 85 maintenance men. The Forest Service recruited 52 men from Stockton. The resident population in the valley added 153, drafted from the Yosemite Park and Curry Company, the state fish hatchery, the post office, Standard Oil

FIGURE 14 Rancheria Mountain fire (1948).

FIGURE 15 One of eight fire camps for the Rancheria Mountain fire (1948).

stations, the City of San Francisco (because of its link to Hetch Hetchy), and the Davis Lumber Company; Fort Ord added 220 soldiers. Yellowstone, Glacier, and Mount Rainier National Parks furnished supervisory overhead. Communication was spotty: a telephone line from Harden Lake to Pate Valley had ceased to be maintained when the CCC left, and the park's radio system was "antiquated and woefully inadequate." Logistics ("service of supply") had to follow the same routes as the crews. A 38-mile long road led from the valley to O'Shaughessey Dam; boats then took men and matériel two miles to the mouth of Tiltill Creek ("called Omaha Beach during the fire") from where it was 4.5 miles by trail to Rancheria camp, 14 miles to the Pleasant Valley camp, and another 4 miles to the Junction camp, all serviced by "78 horses and mules" organized into pack trains to supply 8 camps. In the end, the fire burned 11,800 acres. There were no injuries.[22]

* * *

But while fire had always mattered to the NPS, visitors mattered more, and tourism boomed in the postwar era. In response, the Park Service successfully lobbied for a 10-year, agency-specific echo to the New Deal's Emergency Conservation Work. Mission 66 extended from 1956 to 1966, the 50th anniversary of the agency's organic act. It targeted infrastructure, from visitor centers to offices and employee housing. Similarly, the surge of tourists made preventing human fire starts through visitor contact and campgrounds the focus of understaffed fire programs. Massive planning and public works—the Park Service had little interest or bandwidth to reconsider its fire program. The dictum established by Coffman persisted: firefighting took precedence over all other activities, save the safety and safeguarding of human life.[23]

Less expectedly, the heavy hand of the Forest Service was poised to lift. In 1960 Herbert Kaufman published a celebrated study in public administration, *The Forest Ranger*, in which he identified the agency as a paragon of governing bureaucracy. Fifty years after the Big Blowup, the Forest Service had seemingly met its fire protection goal: it controlled almost all firefighting resources; had near monopoly over any research into fire; oversaw fire prevention, with Smokey Bear as a universally recognized emblem; furnished the matrix for state and other federal fire programs; spoke with the putative authority of academic forestry; and determined policy. It was, by any meaningful definition, a hegemon. Fifty years later Francis Fukuyama identified the Forest Service as the epitome of dysfunctional democracy. Even by the 1970s it was no longer in a position to impose its standards on others or act as a flywheel within the dispersed matrix of the American fire establishment.[24]

The Department of the Interior (DOI) began to create an institutional rival. In Alaska, in the Great Basin, through the emergent Bureau of Land Management (BLM), Interior created an alternative agency for fire. In fact, funding for fire was a principal mechanism behind the BLM's build-out. Added together, the BLM, the National Park Service, and the Bureau of Indian Affairs gave the Interior Department a heft in fire it had never had before. In 1969 the BLM had become strong enough to negotiate a dual-administered national center for coordinating fire support. The Boise Interagency Fire Center (later, National Interagency Fire Center) announced a challenger if not yet a counterweight to the Forest Service. But that competition did not extend to policy. Many of the BLM's managers had learned their craft with the USFS, and fire suppression was what granted access to the money the agency needed to expand its administration. A revolution in policy had to wait.[25]

Yosemite was too visible and vital a park not to share national trends, and too autonomous not to have its own idiographic history. For a century it had experienced regime change as the Miwok left, as shepherds relocated fires from valley and groves to the high country, as the army suppressed fires and fire starters, as the occasional big fire galvanized resolve, as the CCC strengthened an infrastructure and staffed fire lines, as postwar investments doubled down on fire control. By means both overt and subliminal fire had receded. A big fire broke out perhaps once a decade, not something that could be planned for or that merited stripping resources away from the remorseless crush of visitors and their caterers.

The fire scene at Yosemite had deviated far from the landscapes prior to the Mariposa Battalion's entry into the valley. The era of the 10 a.m. policy—however monolithic and unmovable the doctrine appeared to critics—informed the park for roughly 30 years. Only during the CCC era had the park sufficient manpower, equipment, and purpose to apply it fully.

• • •

The park had trodden a path that periodically forked between fighting fires and lighting them. At each junction it had chosen (or let happen by default) tighter fire control, and the path eventually punched so deeply into the backcountry that no other option seemed possible. The threats worsened; more suppression seemed the only response. The park could not go back to the time of the Miwok, or of Commissioner William Miller, or of Captain Gale, or of Sequoia Superintendent John White, and choose systematic fire lighting over systematic fire fighting. At issue was not just the path chosen but the growing momentum and weight of all the choices in staffing, equipment, training, public expectations, and agency culture that discouraged it from not just walking back to an earlier trailhead and trying again but from finding a way to move from the current position to something like what might have happened had that other path been taken.

In the 1960s, however, Yosemite decided to do just that.

CHAPTER 5

The Trek

The Path to Illilouette

Along Horizon Ridge, roughly a dozen miles before it reaches Glacier Point, the road forks. The paved road abruptly shifts 90 degrees, elbowing northward, while any easterly traffic must follow a foot trail that splits off into the Yosemite wilderness and the broad basin of Illilouette Creek.

The trekking party has scaled down from 15 to 12. Life has intervened. Broken cars, calls from higher-ups, the relentless crises of administration have pared down the ideal roster to a working one. The superintendent, her deputy, and the chief ranger, all have other, unyielding claims for their attention. They have no time to contemplate: they will deal with the fires as they arise. Instead, a former wilderness manager and another academic agree to join for a day.

Packers unload horses and mules to carry the heavy gear and metal bear boxes. There are three strings in all, enough to haul a duffel from each trekker along with the collective camp gear. A map and verbal instructions confirm the drop location. The pack string leads, since the trekkers will halt for short talks. Allowing time for the mule-churned dust to settle, the party sets off for Mono Meadows and the elusive, some thought hallucinatory, vision of fire restored that had risen over the cliffs 50 years earlier.

The morning sun glistens. Yosemite's two active fires along the Tioga Road—the County Line and Lukens, begun on June 28—continue to creep and sweep within the biotic box established by last year's burns. To the south two lightning fires in Sequoia-Kings Canyon, the Colony and the Paradise fires, set three days before, are strengthening. Yosemite answers their first call for assistance.

• • •

A crude oval, the shape of the Illilouette basin mimics the shape of Yosemite overall. Half Dome, Mount Starr King, the Clark Range, Buena Vista Crest, Horse Ridge, and Horizon Ridge rim the basin, loosely isolating it into a biotic island. Illilouette Creek itself flows northward until it merges with the Merced River at Half Dome shortly before spilling into the valley as Illilouette Falls. A fugue of stone and flowing water, dappled and fringed with forest and shrubland, defines the character of its vistas.

The scene within the watershed of the Illilouette proposes an alternative, for here the animating flow takes the form of fire, not water. Since the Big Blowup of 1910 effectively announced the onset of America's modern era of wildland fire, the country had spent half a century trying to take all fire out of its landscapes; it would spend the next half century trying to put good fire back in.

That inflection point occurred in the 1960s and 1970s and it had two poles that like a bar magnet held the national fire scene in its force field. One pole resided in Florida. The other lay in California, anchored in Sequoia-Kings Canyon National Park and Yosemite. The Florida insurgents, operating out of Tall Timbers Research Station, spearheaded an accommodation to prescribed fire. They embraced private as well as public lands, wanted a restoration of local fire customs, and promoted a landowner's right to burn. Their poster child was the longleaf pine, the fire-hardy master of the coastal plains (and the Southeast's cognate to the West's ponderosa pine). The California protest focused mostly on public land and wilderness, although echoes of the light burning controversy lingered like spores ready to sprout when conditions favored. Their poster child was the giant sequoia.

The paved road takes tourists to Glacier Point, its sweeping panorama of an ice-sculpted vista, and for much of the park's history, the site for its fire-as-spectacle evening Firefall. The trail takes the trekkers to Illilouette, a patchy quilt of Tuolumne granodiorite and mixed-conifer forest, and for the past 50 years, a landscape deftly managed to restore natural fire. Illilouette is to Yosemite's wilderness what the Mariposa Grove is to its clusters of giant sequoias. Together with grove and valley, it is one of the three apexes where the fire revolution came to Yosemite.

• • •

It was both a typical revolution, sparked by values, rising expectations, and an altered way of viewing the world, and an odd revolution, led by eccentric elders, not by the Sixties' fabled youthquake. During its opening salvos in 1962, Ed Komarek was 53, Harold Biswell 56, Harold Weaver 58, and the revolution's patriarch, Herbert Stoddard, 73. Their ambition was restorative: putting fire back on a land that needed it, putting the torch back in the hands of traditional users, letting lightning work its fire-kindling marvels in the backcountry. Youngsters would inherit the charge to implement the new vision.

Its institutional markers arrived in steady succession, galvanized the same year Rachel Carson rallied environmental enthusiasms and alarms with *Silent Spring*. In 1962 the Tall Timbers Research Station, a privately endowed facility north of Tallahassee, launched a series of annual fire ecology conferences that proposed a countervoice to a fire establishment dominated by the Forest Service. That same year, at the Allison Prairie Preserve outside Minneapolis, the Nature Conservancy, another nongovernmental organization, conducted its first prescribed fire. To complete the outburst Interior Secretary Stewart Udall commissioned three reports, each of which affected Yosemite. The Outdoor Recreation Resources Review Commission (ORRRC) published its *National Recreation Survey* in 1962. The National Academy of Sciences Advisory Committee on National Park Service Research, chaired by William Robbins, released its report on scientific research in the national parks in 1963. And not least in 1963 the Advisory Board on Wildlife Management of the National Parks, created in response to a festering crisis with elk at Yellowstone, and headed by Starker Leopold, a professor at the University of California–Berkeley, published its conclusions. All three commissions were chaired by faculty from the University of California–Berkeley, all in some way touched on that particular portion of the backcountry known as wilderness, and all looked to Yosemite.

The ORRRC included wilderness among outdoor recreation sites, although it confessed "that unmodified natural conditions do not exist in any extensive area in the continental United States." It identified as a disruptor of "special importance" the pervasive influence of "fire suppression," and as a "countermeasure" to the "disappearance of important wilderness values," it suggested "different and perhaps less rigorous fire protection policies for portions of some wilderness areas." The Robbins Report noted the urgency for a scientific basis for managing natural areas—a capacity the NPS lacked—and the value of relatively unmodified lands as research sites. But the bombshell was the Leopold Report. The Advisory Board on Wildlife Management urged the National Park Service to turn away from its Mission 66 agenda and think more deeply about managing its natural estate, not as landscape architecture but as natural systems behaving more like they had in presettlement times. It singled out the forests of the Sierra Nevada as an exemplar of a landscape overturned by the systematic removal of fire.[1]

Clearly, preservation was an idea whose time had come. In 1964 the Wilderness Act created a new category of federal land for which natural processes should be allowed to proceed untrammeled. No one imagined fire suppression, especially mechanized fire protection, continuing unchanged in wilderness; but neither did anyone advocate standing aside and letting natural fire creep and sweep unchecked; and almost no one thought that anthropogenic fire had any place in the formal wild, much less that it might be indispensable in shaping the landscapes being preserved. Mostly, wilderness advocates wanted to keep roads and bulldozers out, and did not want fire control to run engines roughshod over wilderness character. Yet as one fire officer put it, the Wilderness Act made extinguishing a naturally caused fire "practically illegal." The wild belonged in the wild; the conundrum was convincing wildfire to stay in its designated sanctuary.[2]

Meanwhile, other environmental reforms for clean water, clean air, and endangered species added to decision trees. And more subtly, the Civil Rights Act, the Voting Rights Act, and the Immigration and Nationality Act of 1965 pointed to an upheaval in national demographics and social relations, one that left forestry, almost exclusively the domain of white men with common training and temperament, increasingly isolated from national sentiments. Whatever their policy the fire agencies would have a different workforce to apply it.

• • •

The Leopold Report's charge was wildlife, but wildlife relied on habitat, and many habitats reflected their fire history. In a long passage oft-cited at the time, the report explained how fire exclusion had harmed the Sierra Nevada forests. "When the forty-niners poured over the Sierra Nevada into California, those that kept diaries spoke almost to a man of the wide-spaced columns of mature trees that grew on the lower western slope in gigantic magnificence. The ground was a grass parkland, in spring time carpeted with wildflowers. Deer and bears were abundant." A century later, "much of the west slope is a dog-hair thicket of young pines, white fir, incense cedar, and mature brush—a direct function of overprotection from natural ground fires," and the Sierra's four national parks—Lassen, Yosemite, Sequoia, and Kings Canyon—had thickets "even more impenetrable than elsewhere." Such feral woods were a tinder box ready to explode, even as they made poor habitat for wildlife and wildflowers, and "to some at least the vegetative tangle is depressing, not uplifting."[3]

The report then asked whether it was possible "that some primitive forest could be restored, at least on a local scale? And if so, how?" The committee confessed, "We cannot offer an answer." But surely the traumas caused by fire's removal might be healed by fire's restoration. "Of the various methods of manipulating vegetation," the report asserted, "the controlled use of fire is the most 'natural' and much the cheapest and easiest to apply." It cited examples from Everglades to the Midwest prairie to East African savannas. Such fires might burn piles or through grassy understories or even by "periodic holocausts." The particular mix depended on local conditions. What was certain was that "rebuilding damaged biotas" would "not be done by passive protection alone." Halting fire's suppression might not be enough. Active burning would likely be necessary.[4]

The report came like a thunderclap. Secretary Udall instructed the NPS to reconsider policies; and in a shockingly short period of time the old edifice crumbled. Over the winter of 1967–68 the agency issued new administrative guidelines for its natural parks, recreational areas, and historic sites. The Green Book, as the manual for natural areas became known, recanted the 10 a.m. policy and urged that natural fire be restored, or where necessary, prescribed fire substituted until such time as natural fire might be possible. Over the next decade the Forest Service, through a series of stutter steps, moved toward a similar policy, formally issued in 1978. In its revision of the Green Book that year, the Park Service affirmed that many fires are "natural phenomena which must be permitted to continue to influence the ecosystem if truly natural systems are to be perpetuated." They were to landscapes what predators were to wildlife, not simply charismatic symbols but processes fundamental to ecological integrity. The two agencies began expanding cooperative agreements to incorporate the possibility of desirable fires crossing boundaries, especially when wilderness also spanned those borders. The free-burning fire joined the free-ranging wolf as an ineffable token of the wild.[5]

The reformation was sudden and breathtaking, and to the young, exhilarating. Bob Barbee, Yosemite's first dedicated resource manager, recalled "there was a lot of putting out all fires. I mean, just all kinds of things that were warped, as we now look at it. But Leopold and that committee took a look at this and they came up with a report that just shook the foundation of the National Park Service. And it became a manifesto, so to speak, for change." Then he posed the critical question: "Who's going to change?" Who would make change? And who would be affected by it?[6]

• • •

The origin story of the fire revolution has been told and retold until it is worn as smooth as river cobbles. There were a few renegade foresters like H. H. Chapman at Yale University, studying southern pines, and Harold Weaver of the Bureau of Indian Affairs, doing the same for the western yellow pine. But most of the insurgents were wildlife people, and though scientists, science did not really lead the charge against the bureaucratic barricades. Research followed policy; it didn't inform it. The basics had been known for decades.

What changed was that influencers saw the same scene differently: it was a gestalt switch, like two facing heads that suddenly appear as vases. During the light-burning controversy, the two sides saw the identical scene in opposing ways. During the fire revolution, the same gestalt process happened again. The founders reimagined fire not as something that landscapes adapted to simply in self-defense, but as something necessary to their metabolism. It wasn't enough to stop suppressing fires. They had to be restored.

Starker Leopold deftly illustrates how the switch in perception could toggle into an inverted understanding of fire. When he joined his father in 1938 for a trip to the Sierra Tarahumara in northern Mexico, they found a landscape awash with routine fire, a fusion of lightning and anthropogenic burning. "The thing that astonished me about that Gavilán country was that it burned every year—fires just kept running through there." They watched, stunned, while their Mormon guides casually tossed matches to the ground while they rode into the mountains. Starker confessed that he "never saw a healthier piece of real estate."[7]

At night, over campfires, they discussed the scene. Aldo observed that Mexico had contained its cattle but left its fires to spread, and had exemplary landscapes. The United States had suppressed fire but left cattle to overrun everything, and had trashed landscapes. "It began to dawn on me," Starker recalled, "that fire was a perfectly normal part of that sort of semi-arid country, and might even be an essential part of it. And Dad, who had been brought up in the Forest Service with the tradition always against fire, he began to wonder too." But Aldo, one of the founders of the Wilderness Society three years earlier, didn't make the leap from land that could accept fire to land that needed fire. Over time, Starker did.[8]

Ten years later, a few months after his father had passed away, Starker revisited the Rio Gavilán, now opened to logging and grazing; fires were scarcer and the land was damaged. "I came home with some ideas about fire that I'd never had before. And when I started looking around California at some of the situations that you could see right from the highway—including our own national parks, Yosemite—I was struck with how prevention of fire was creating tremendous fire hazards in the thick growth of white fir and incense cedar and other stuff. I began to wonder if this is really the way to manage this kind of country, completely excluding fire, which had been a natural part of the countryside."[9]

Such notions, further marinated over the years, entered his eponymous report. Some fire-savvy colleagues at Berkeley undoubtedly helped crystallize those inchoate notions into a philosophy of nature management. For a dissertation advisor he had Carl Sauer, a historical geographer and one of the few scholars who had seriously considered the meaning of humanity's ancient alliance with fire. And for a model of moving ideas into practice, he had a range professor in the School of Forestry, Harold Biswell, who made restoring fire to California's landscapes a personal mission.[10]

Still, he remained rooted to the university. Even when he agreed to head the embryonic science program conjured into existence for

FIGURE 16 Starker Leopold, packing a deer kill during his 1948 return excursion to the Rio Gavilán. Fourteen years later he would chair the Advisory Board on Wildlife Management of the National Parks, and his experiences in Chihuahua would influence his perception of the Sierra Nevada and the ecological role of fire.

the Park Service by the Robbins Report, he did so from Berkeley. The arrangement lasted only a year. His true role was that of professor and oracle.

• • •

The proof-by-doing task fell to Harold Biswell. Born in 1905 Biswell grew up in Nebraska, whose grasslands had also nurtured such botanical luminaries as Charles Bessey, Frederic Clements, and John Weaver. He joined the Forest Service's California Forest and

Range Experiment Station in 1930 as a range scientist; transferred to the Southeast Experiment Station in Asheville, North Carolina, in 1940; and returned to Berkeley as a professor of range science within the Forestry Department in 1947. He remained at Berkeley through his retirement in 1973 as an extension agent.

When he first arrived in California, he held the presumptions and prejudices against fire typical of the agency; when he returned 17 years later, he was a vocal advocate for deliberate burning. His conversion experience occurred while in the Southeast where he saw traditional burning and the compromise crafted by a station colleague, Raymond Conarro, that became known as prescribed fire. He brought those experiences to California, and revived light burning, still latent among ranchers, tribes, and wildlife enthusiasts, now reincarnated. In 1945 California had arranged a permit system by which ranchers might burn to control brush and stimulate forage, which Biswell tapped into (like most California efforts to promote burning, the program faded, passing roughly at the time of the fire revolution). In 1951 he translated the technique for ponderosa pine, with the Teaford Forest in the Central Sierras and Hoberg's Resort in Lake County as showcases, and in 1964 to Whitaker's Forest, a University of California field station, at Redwood Mountain, along the west boundary of Sequoia-Kings Canyon.[11]

When the Leopold Report was issued, Harold Biswell—Doc to his students, Harry the Torch to his critics—was 58; 63 when the NPS released the Green Book; and 68 when he retired and both Sequoia-Kings Canyon and Yosemite National Parks had functioning fire restoration programs. He continued to consult for agencies (including the National Park Service and the California State Parks), to inspire acolytes, and to promote prescribed fire, typically with field demonstrations, often in the face of vituperative criticism and professional ostracism by the fire establishment. When the University of California Press asked him to write a book about his experiences, he wrote not an autobiography—this was not about himself as herald and overlooked genius—but a congenial handbook, told

FIGURE 17 Small fires, Big Trees. Three of the fire insurgents at Yosemite: [back row] Bob Barbee (left), Harold Biswell (second from left), Jan van Wagtendonk (second from right).

in colloquial language. He was 84 when *Prescribed Burning in California Wildlands Vegetation Management* was published. He died three years later. More than the book, his legacy persisted through his students, the demo plots at Redwood Mountain, and the institutionalization of his thoughts in practice.

Controlled burning had been around far longer than the Sierra parks, and arguments about it had coevolved with park administration. What mattered was the chrysalis among ideas and the ability to embed them into agencies. Wildland fire didn't need more prophets in the wilderness. It needed ways to translate philosophy into practice, practice into manuals, and manuals into organization charts and institutions. Harold Biswell made that happen. He had institutional support, through his tenured professorship, that allowed him to pursue his campaign; had students, who could formalize and propagate the concept; had access to the Whitaker's Forest Cooperative Natural Research Station jointly operated by the University of California–Berkeley, the NPS experimental for-

FIGURE 18 Lighting the revolution: Harold Biswell (left) and Jan van Wagtendonk take a breather during a 1970 prescribed burn.

est at Redwood Mountain outside Sequoia-Kings Canyon, where demonstration trials were possible, and the California Department of Parks and Recreation; and had in the giant sequoia a charismatic megaflora that the reformists used as a poster child. Unsurprisingly, he and Starker Leopold knew each other and cotaught a seminar. He hosted the Seventh Tall Timbers Fire Ecology Conference when it met in California in 1967. He had picked up the fire stick by its other end and waved it for all to see.

Mostly he had those students—his, and those like Bruce Kilgore he shared with Leopold. They were the ones who found, invented, improvised, nurtured, coaxed, conjured, and otherwise concocted a fulcrum that allowed Biswell's drip torch to become an Archimedean lever that could move Yosemite. They wrote textbooks and manuals, they devised prescriptions, they engaged in the bureaucratic trenches where ideas would either prevail or falter. The redoubtable "Dr. Biswell" had, Moses-like, led them to a promised land. They were the ones who would subdue it.

Sequoia-Kings Canyon and Yosemite soon made a tag team that established an exemplar for the West. In 1968 Sequoia-Kings had "let burns" in the backcountry and a prescribed burn around the groves. With Biswell as a conduit and mentor, Bob Barbee, Yosemite's first dedicated resource manager, carried the concept north. Smart, athletic, willing to take acceptable risks, he seemed to look on those who were challenging fire tradition as rock climbers did the clique at Camp 4 that was revolutionizing goals and techniques. At Biswell's invitation, Harold Weaver showed up to offer his imprimatur. When the Forest Service sought to introduce natural fire into the Selway-Bitterroot Wilderness, it sent two planners to observe the Sierra scene.

Those were heady years in which solutions seemed not only possible but relatively obvious. If fire suppression had unwisely eliminated fire, then stop suppressing. Where the ecological damage was severe, intervene with slashing and burning until conditions could allow natural fire to reclaim its rightful place. The mandate was, in an odd way, the National Park Service's version of *Star Trek*'s Prime Directive (the TV series ran from 1966 to 1969). Don't intervene, but if intervention happens, take steps to return to the prior conditions and then stand aside.

• • •

At Mono Meadow the trek pauses, as the group moves to the side of the trail to let the long pack string pass on its return trip to Horizon Ridge. The site is shady and still wet even at the end of California's endless summer. The pause provides a useful opportunity for impromptu discussion. This is, after all, a Park Service nature walk, and it would be incomplete without interpretive guides.

Crystal Kolden, a pyrogeographer from the University of California–Merced, talks about fire and California with special emphasis on native fire history and cultural burning. For the future to succeed it will have to resemble that former past. Scott Stephens, a fire scientist at UC–Berkeley, then gives a short preamble on the 20 years he and his lab have devoted to the Illilouette project. The next day's hike will venture deeper into some of the prime sites, so he defers much elaboration. But he notes a powerful puzzle that bubbled up from their reconstruction of fire history. The chronicle of fires as inscribed in scarred trees abruptly breaks around 1875 and does not return until the Starr King fire in 1974. For a century fires simply vanished from the record. The reason why is not understood.

Crystal notes that the dates correspond with the final expulsion of the Miwok people. Mark Fincher, the park's long-tenured wilderness planner, observes that cattle may have entered the basin, competing with fire to consume the grass and shrubs. The early development of Glacier Point provided for rapid detection of any smokes and their possible quick suppression. Whatever the ultimate conclusion over causes, it is clear that Yosemite's fire history is inextricably connected with people. Two lightning fires started earlier in the summer in Illilouette; fire crews extinguished both.

• • •

Under the shock of the Green Book, Yosemite had to reconsider its entrenched habits. Perhaps nothing felt the new regime so fully as fire since it alone touched all the other park concerns. The old debates about fire in valley and grove revived, and were joined, thanks to the Leopold Report and Wilderness Act, with a third apex to complete its fire triangle: what had traditionally been seen as backcountry acquired a new significance as the Wild. The valley could be imagined as landscape architecture, a practice the Park Service knew something about. The grove could be regarded as a historic site, though complicated by its biological materials. The wild had, for the agency, no precedent for managing other than to exclude fires, a practice it now had to reverse.

The Valley

Some patchy burning continued in the valley up to 1930. The Ahwahneechee provided plenty of precedent, those newcomers in uniform who had grown up on farms were accustomed to burning fields and pasture, and the consequences of not burning were dramatic and obvious to anyone not educated to the point that their minds could not picture what their eyes saw. But Indigenous fires supported the Ahwahneechee economy; any successor versions had to coexist with, if not support, an economy of cars, campgrounds, and concessioners. Nor had annual firing evolved alongside the hydrology that the newcomers reconfigured by destroying brush and moraine dams and draining wetlands (to control the mosquito population). The lingering fires could not prevent further reforesting, nor could they drive out what had rooted.

Even after the Park Service assumed administration, some selective burning of meadows had continued. Chief Ranger Forrest Townsley (ca. 1914) claimed it was rather common. (Among the postburn regrowth were masses of evening primroses, until the valley deer population grew too large.) Other accounts speak of burns in 1919, 1920, 1921, and 1930—and these were only those episodes that left records. "Such burnings," summarized Emil Ernst, "were always localized to particular meadows and not allowed to become widespread" as had been the practice by the Ahwahneechee, cattlemen, and shepherds. In brief, despite official policy, an informal economy of fire had endured, a kind of ecological black market.[12]

Coffman and the CCC ended any sub-rosa burning, but they could not stop the valley from overflowing with trees, now growing in number if not in stature. As the valley's ecology veered ever further from historic norms, so its vistas—the reason for visitation—became obscured. The NPS had long associations with landscape architecture, but given the high visibility of the valley, removing trees and burning them to black stumps was sure to cast sparks of public protest. Instead, year after year parties went out and mechanically clipped back critical sites. Then, with Biswell as a mentor,

Bob Barbee managed to burn off the El Cap Meadow and in 45 minutes killed every young pine. Everything but the trees sprouted back the next spring. While a nifty proof of concept, the program did not expand beyond demos.[13]

The most pressing crisis, however, was not the number of trees but the number of visitors. The valley was swamped by people and cars. The agency's traditional response had always been to enlarge, carefully, facilities to accommodate them. Until it had a master plan for the valley, however, restoring the old grasslands was hypothetical. A few experimental burns were possible; wholesale restoration had to wait. Until then fire management meant structure protection.

The park's primary gesture was symbolic: it finally banned the phony Firefall. The public argument was informed by the vision of a "primitive America" urged by the Leopold Report. The practical purpose had to do with the massive congestion, litter, and general unruliness the spectacle encouraged, as though embers rushing down granite was Yosemite's version of a poorly policed rock concert. (Woodstock occurred the next year.) Worse, on the Fourth of July, 1970, as rangers attempted to disperse what they regarded as an illegal gathering, a riot resulted at Stoneman Meadow, the rangers were routed, and managing visitors, not fire, became the obsession of the valley's administrators. The riot's aftermath sent the park's senior staffers elsewhere, a lesson not lost on their replacements.[14]

The park could banish firefalls. It could not as readily put real fire into the valley.

The Grove

The same problems afflicted the Big Trees, which were being trampled by people and overgrown by a relentless encroachment of white fir and incense cedar that not only obscured the giant sequoias from view and prevented seedlings amid their gloom, but threatened even the giants with fires that could leap into their crowns. In Muir's day a fire spreading into the grove would promptly go to ground, easily contained by swatting with a burlap bag. By 1963 a fire could blast through the canopies and likely immolate even mature sequoias. Not least, the giants were not regenerating.

Forest invasion was an old concern. Again and again, with little fanfare, guardians beginning with Galen Clark had cleared away both the in-filling forest and the stubborn shrubs that ringed the bases of the Big Trees. With CCC labor, by October 1933, over 3,000 trees had been felled and hauled away. The pressures were endless, like a river pressing against a levee. Rangers could erect fences to prevent trampling around the root collars, but the insistent woods moved in where tourists couldn't. More worryingly, no young sequoias were regenerating to replace the aged ones. The biology behind these concerns was unknown.[15]

In 1956 Richard Hartesveldt at San Jose State University began research toward a doctoral study on human impacts in the Mariposa Grove. A photograph taken in 1890 showed enormous boles standing like megalithic monuments amid a grassy understory with

FIGURE 19 Mariposa Grove (1890).

FIGURE 20 Mariposa Grove, same perspective (1970). Despite being mechanically cleared repeatedly, once comprehensively, the grove is overgrown with white fir and incense cedar.

the occasional fir or cedar in the background. A rephotograph of the same scene in 1970 showed a dense forest among which here and there an enormous trunk was visible above the impenetrable canopy. The situation was indefensible.

In 1962, with his dissertation completed and with support from the Sequoia Natural History Association, the same year Secretary of the Interior Udall commissioned the Leopold and Robbins Reports—Hartesveldt transferred his attention to the groves at Sequoia-Kings Canyon National Parks. For a charismatic tree responsible for three of America's first four national parks, shockingly little was known about its basic biology. A year later Hartesveldt was joined by Thomas Harvey, and together they focused on fire and sequoia regeneration. Later, other researchers joined to incorporate fauna. The Park Service added funding through research contracts;

the Whitaker's Forest Cooperative Natural Research Station under UC–Berkeley assisted, creating a critical mass around Redwood Mountain. Interim reports tracked their discoveries in 1965, 1967, and 1970; a final report was submitted in 1971, and published in 1975, when Hartesveldt unexpectedly died.[16]

What his group found for the Sequoia-Kings Canyon groves applied to those at Yosemite, which included not only the celebrated Mariposa Grove but also the Tuolumne and Merced Groves. Fire and sequoia were intimately linked: repeated surface fires kept the revanchist white fir and incense cedar at bay; patchy hot fires created the bare soil conditions needed for sequoia seeds to germinate. The survival of the Big Trees depended on the right medley of fire. Their impervious bark protected their trunks; immense fire scars demonstrated their capacity to endure fires that could carve out cavities, and repeatedly return; but these were traits helpful to overcome the endless fires that, over hundreds of years, had to occur until the right circumstances of fire and seed sparked regeneration. In the bland language of the study group, "the role of fire in sequoia groves is now seen to include both fuel reduction and enhancement of giant sequoia regeneration."[17]

It was not enough to remove the invasive trees, as fuel; the site needed the ecological alchemy of fire. With Hartesveldt's group bestowing scientific validation, and the Green Book granting policy approval, Biswell and his students busily created demonstration plots. The Sierra parks crafted fire management plans that sought to rehabilitate the grove, an example that others emulated. In 1975 the Forest Service set a test fire in the Nelder Grove near Yosemite and the California Department of Parks and Recreation burned in the South Calaveras Grove.

But questions persisted, equally scientific and managerial. The Leopold Report had established the landscape of the presettlement era as a restore point and resulting policies assumed that, once reset, the former landscape would self-regulate itself. The sequoias, however, preserved a past far older than the gold rush. Amid the early glimmers of climate change, the U.S. Geological Survey contracted with the Laboratory of Tree-Ring Research at the University of Arizona to use the millennial chronicles of sequoias to reconstruct fire regimes across the entire duration of the extant giant sequoias. It was hoped that the *longue durée* fire record might provide guidance on how fire actually worked in the groves and how it might be restored with best effect.[18]

There were, to simplify, two perspectives. One held that the sequoias responded primarily to the particularities of place; they were after all very localized into clumps. Their fire history would reflect an internal rhythm peculiar to those select sites, while the giants—botanical authorities in taking the long view—shrugged off climatic flickerings as they did fire and beetles. Management need only recreate the fundamentals and let the groves' intrinsic ecological clockwork run. The other view held that sequoia history, like all history, was idiographic, the outcome of "unique, aperiodic external forces" that were "unique at any given point in time" and "basically un-repeatable." Fire scars, for example, tracked the long climatic wave of the Medieval Warm period. Nor was it all climate. Intriguingly, the Mariposa Grove differed from the trends apparent in cognate sites by revealing a more profuse record of fires. People, too, it seems, had their long waves of coming and going.[19]

If the park wanted sequoias—if salvaging sequoias amid the havoc of the Anthropocene was even possible—management had to be nimble, responsive, and endless. Whatever the strategy chosen, and whatever the ultimate fate of the Big Trees, the researchers believed that "the reintroduction of fire to the sequoia groves is both an ecological imperative and an opportunity to mitigate negative impacts of human-caused changes in the environment." The trick was to have fires cool enough and frequent enough to cleanse the revanchist intruders, yet have patches that burned hot enough to expose mineral soil stripped even of grass and ready for seeds.

Before anyone could put a torch into the grove, the metastasizing forest of encroaching firs and cedars had to be excised. Bob Barbee recalled how "we went out to Wawona with Dr. Biswell and we wrote prescriptions. And then I went out to Mariposa Grove and we looked at that, and said, 'There's no way we can start burning this right now. . . .' I mean, it's a jungle, a jungle." They had "to do something," but they couldn't burn it because "the trees are too big and, you know, you could have a holocaust here, kill all the trees." So they cut.[20]

"We cut tens of thousands of trees in the Mariposa Grove," Barbee recollected. "No science to it. It was all done with the eyes of—I thought—artistically, so to speak. There wasn't any way to do it scientifically, at least not then." They left a few white firs (Barbee identified them with flagging tape, so they would be spared). His crews stacked and piled the slash. "And then when it rained we would burn them. That's the only way we could try to set back the sequoia groves." Broadcast maintenance burning would follow.[21]

In principle, the cutting was a prelude to burning. Fire was more ecologically benign and unlike cutting it created conditions that promoted regeneration. But it was also less tamed, and it left scorch marks and inscribed catfaces on the trunks that were visible in ways stumps flushed to ground level were not. An axe opened vistas to the Big Trees: visitors saw the trees, not the stumps. When a burn opened vistas, visitors could see the marks of the fire. Not everyone was pleased.

What happens in one park can affect them all, especially when they share a common geography or task. Those unhappy with a program speak out; those content with it generally do not. That's true even when the criticism comes internally. Some fires burned hotter than intended and with unexpected results. In 1982 dissent at Sequoia-Kings Canyon went public, but the 1985 Broken Arrow prescribed fire in Giant Forest aroused a local resident and former park seasonal, Eric Barnes, to object sufficiently that the Park Service commissioned a review of its program. The committee, headed by Norm Christensen, a fire ecologist at Duke University, issued its report in 1987. A year later Yellowstone National Park's prescribed natural fire (PNF) program blew up to obsessive media attention, and PNFs were banned until parks revised their fire plans according to national guidelines. Norm Christensen was again enlisted to chair a review committee.[22]

In the Sierra parks competing voices argued over a dog-hair thicket of controversies that ranged from philosophy to policy to practice. Building on the Leopold Report, the Christensen committee came down firmly for the preservation of ecological processes over the maintenance of a static landscape, which in any event was impossible. But the groves were special, the Mariposa Grove a biotic Parthenon. The committee accepted the park's designation of them as "showcases." If it was inappropriate to project strategies suitable to the groves outward across the backcountry, so it was equally inappropriate to project backcountry fire management onto the groves. Regeneration was a long-term issue; fire threats from thickening woods, an immediate one.

The committee accepted that the historic fire regime had included Indigenous burning, that fire exclusion had allowed white fir and incense cedar to clog views and fluff up fuels to dangerous levels, that aesthetics did matter. Accordingly, it did not believe a lightning-only ignition regime was warranted, it encouraged that fuels be cleaned up, and it urged that the park fire program involve consultation with someone "educated in landscape architecture." In effect, the parks needed multiple fire plans to accommodate their variety of settings.[23]

Apart from the reset, the fire program had plenty of concerns. The institutional arrangement that left fire fighting with the Protection Division and fire lighting with Resource Management could only work if, as with a marriage, all parties valued it equally. But this was a shotgun wedding of unequal partners. Fire fighting had ample money and access to emergency funds as needed. Fire lighting had to budget for each event. Jan van Wagtendonk commented acidly that the arrival of a new fire officer in the Protection Division, committed only to suppression, "set the program back decades." The park soon met the criteria for renewing its operations, but every glitch only seemed to go one way, toward lessening the opportunities for deliberate fire. Prescribed fires ran on checklists, any item of which could cause a burn to be shut down. There was no comparable checklist by which to jumpstart a burn or reinstate a cancelled one.[24]

Even in the groves the program slowed. It had launched amid environmental enthusiasms broadly diffused throughout American society and politics. By the 1980s headwinds replaced that following sea.

The Wild

To Yosemite's classic fire scenes the 1960s had added a third, the wild. What began as backcountry would, by 1984, morph into legal wilderness. This was a lot of land, and a lot of fire to consider. Certainly, there had been plenty in the past without having to take literally observers' endless comments that "all the forest" was burned annually. Since its cavalry days, the park had attempted to exclude fires in the backcountry, whatever their cause. Now, as with valley and grove, it needed to reinstate at least some fraction of those fires that nature kindled.

With the Green Book's sanction, the park began allowing some room for natural fires to roam. This was a novel enterprise: there had been examples of benign neglect, of fires simply ignored or too small to fret over. But a deliberate exercise to promote backcountry fires was an innovation. It couldn't be solved by cutting out unwanted trees, and prescribed fire seemed—at a minimum—a violation of a wilderness ethos. The public might tolerate felling to improve a view of Half Dome or Yosemite Falls, or to clear biotic clutter around the grove. It was as unlikely to applaud rangers lighting up the backcountry as to approve culling elk by organized hunts. The best option was to find ways to allow natural fire to live out its natural life.

The solution of choice, the prescribed natural fire, was to critics, a contradiction; and to sympathizers, only a paradox. It certainly improved on the impromptu "let burn" label that accompanied the earliest fire reports. It brought to the wild the same kind of legal authority and operational constraints that applied to prescribed fire in valley and grove. Like so much of fire the outcome was obvious in principle, and maddening in practice.

The Park Service compounded intellectual complexity with an organizational arrangement that split the two practices, fighting and lighting, into two distinct administrative divisions. Fire suppression remained with the Protection Division, increasingly dominated by law enforcement officers (the Stoneman Meadow riot reinforced the trend). Fire restoration was shunted into Resource Management, a new entity trying to find its footing within the park's hierarchy and traditional career paths. What should have been a unitary program adept at all the practices required to manage fire was artificially sundered. There were differences, too, in career options. Rangering had long served as the fount from which the NPS drew its managerial cadre. The ranger required no technical qualifications; by training and temperament, rangers were generalists, grounded in visitor services. Resource managers came with formal training in natural science, or (another fallout from the Leopold Report) were sent to special schools for master's degrees (Barbee went to Colorado State, graduating in 1968).

The administrative guidelines made prescribed fire and natural fire acceptable actions; it did not make them preferred solutions, or mandated requirements. So, in addition to distinctions in status, the twin demands of a fire program were also mismatched by operational capabilities. Fire fighting had a budget plus access to inexhaustible emergency funds. Fire lighting had a budget, and each project—even fires monitored in the backcountry and burning for months—had to stay within it.

It was an illogical arrangement, only possible from a genius or a madman. Part of its intent was to protect fire restoration from being browbeaten by traditional fire suppression. Fire management was a start-up, more nimble and better oriented to new visions, not encumbered by old conventions (Yosemite was notoriously prone to entrenched habits). In practice, however, reinstating fire became the poor step-child. No one would intervene to stop fire suppression once begun, but it was easy to halt a prescribed natural fire if conditions jacked up risk. Yet policy would succeed or fail not on its logic or intrinsic value, but on how well it could be implemented. Restoring fire demanded forceful leadership.

That still left the problem of designating places where the new philosophy could be tested. Not every site could freely accept lightning fires, certainly not until more experience accrued. The Illilouette watershed was relatively accessible, large enough to reveal fire dynamics, and even visible from Glacier Point, rendering it amenable to interpretive talks if desired. Rimmed with a granite rampart and peaks, from a fire behavior perspective it was almost ideal for containing fires, and from a research vantage point, an island ecosystem ready for research and modeling. Its weakness was that smoke, like the waters of Illilouette Creek, would nightly flow into the Merced River and down into Yosemite Valley. (As one fire officer put it, "Smoke concerns from fires in the Illilouette Basin are historic"; every major fire "becomes a smoke problem fire.") Fire control was not an issue; smoke control was. However clear in its logic and firmly fixed in policy, fire management became hazy in practice. It knew where it wanted to go. It was unsure how, within a complex bureaucracy and a syncretic society of mixed and often conflicting values, to get there.[25]

In 1972 the Illilouette basin received its first prescribed natural fire.

• • •

Dan worries that dumping our gear and food along the trail might invite vandalism or bears, and he volunteers to hike ahead and guard the cache. His wife, Cindy, a former hotshot and smokejumper, joins

FIGURE 21 Illilouette trek group. (l–r) Steve Pyne; Mark Fincher; Athena Demetry; Joe Meyer; Lacey Hankin; Kelly Singer; Scott Stephens; Chad Anderson; Lisa Hahn; Garrett Dickman; Jeff Webb (upper right). Kneeling in front: Dan Buckley.

him. The rest of the group pauses beside Illilouette Creek. I offer some comments drawn from history, adding a few sidebars to the presentations by Crystal and Scott. The break turns into an impromptu lunch. Then we return to the trail for another mile and a half.

At a shaded fork, where the trail branches to all compass points, we pause while Mark receives a radio call from Dan who says he is unable to locate our cache of gear. He and Cindy have gridded the location without a hint of where our substantial mound of duffels, bear boxes, and stoves might be. Nor, they add, can they find any sign of a pack string recently passing along the trail. We observe the same. A frenzy of radio calls finally connects with the packers. Whether they had misheard the location, or lost the map, or just determined that Illilouette Creek made a more appropriate campsite, they had deposited our gear not 30 yards from where we had halted for lunch.

We reverse and hike back.

CHAPTER 6

The Trek

Burning the Basin

This, the second day, we will trend around the watershed, arcing east toward Mount Starr King, the site of a major fire in 1974, and then toward the Clark Range, where studies have revealed the vital relationship between fire and water. Toward the middle of the basin, Scott lectures at a shallow dome of Tuolumne granodiorite, and then leads us cross-country into a burned site now self-restored to wetlands and aspen.

• • •

Different places at different times in American history have helped inform national policy. The Northern Rockies in the early days of the Forest Service. California in the two postwar eras. Florida during the upwelling of interest in prescribed fire. Sequoia-Kings Canyon and Yosemite for restored fire in the West. Wilderness fire has its own crown jewels: the Selway-Bitterroot, the Gila, Alaska, and not least the backcountry of the Sierra parks. Each agency tends to look to its own example for guidance. Among the national parks, the Illilouette pilot program stood out as a paragon. This is what good fire restoration in wilderness looks like.

Science didn't design the program. Restoring fire to Yosemite's backcountry was urged by the Leopold Report, then implicitly mandated by the 1984 California Wilderness Act that put 95 percent of the park into legal wilderness. In 1970 fire science meant fire behavior. Fire ecology as a scientific discipline hardly existed: it didn't get its own professional society until 2000, and its own journal until 2005. What the UC–Berkeley research group has done is codify the consequences of the Illilouette experiment, offer guidance on techniques, and move into formal scholarship what had been through millennia of human occupancy oral traditions.

Today's excursion will let the trekkers see with their own eyes the places burned and unburned, but they will hear that vision through the voice of modern fire scientists. Projects like Illilouette are as much about personalities as programs.

• • •

By Park Service standards Yosemite in the early 1960s had a stellar program, with abundant wildfires, adequate equipment and staffing, and a long history of fire control. Together with Glacier National Park, it offered a paradigm of serious fire suppression for the agency.

It boasted perhaps 20 percent of the entire NPS fire load. It could rely on an in-park "militia" for backup and had cooperative agreements with the Forest Service and California Department of Forestry. In 1961 it established a helibase at Crane Flat. It assumed additional responsibilities when the park acquired 1,100 acres at El Portal for administrative offices. In 1964 it updated its Fire Control Plan, which noted that the park's "esthetic values alone are invaluable and must be protected from the ravages of wildfire." It had a dedicated fire control officer and an assistant along with a complement of seasonal fire control aids and a forestry crew that could be mobilized for fire duty; few parks were so well endowed. It operated under the dictates of the 10 a.m. policy. The plan affirmed that "fire suppression takes precedence over all other Park activities, for all employees, excepting activities concerned with the saving of human life." All this was standard fare for the time—agency boilerplate. What distinguished Yosemite was its program funding, its public and agency visibility, and the implicit criticism its fire suppression tradition received from the Leopold Report.[1]

A phase change occurred in 1968. The Green Book urged fire management rather than fire control; Barbee was writing a resource management plan that included fire; prescribed fires were kindled in grove and valley; and the fire suppression organization struggled to control the most serious outbreak of wildfire in the park's recorded history which were also "the largest fires this season of any NPS area." The year before, Glacier had grappled with two big fires that went political and prompted a major review. Now it was Yosemite's turn.[2]

The 1968 season followed a hectic 1967 season—"the busiest in years with a total of 157 fires," its fire officer reported. Many fires but not many acres. The 1968 season had 101 fires, but they came one after another without break; racked up more acres; and threatened developed areas. The Union Point, Overlook, Poopernaut, Kibbie, Bishop, and Wawona Dome fires all came up for review. But the prime mover was the Canyon fire that started on September 23, evidently from a faulty automotive exhaust that blew sparks into dry grass on Highway 140 near the Arch Rock Entrance. Within hours the fire was 200 acres, within a day 800. Air tankers, helicopters, engines, bulldozers, and crews worked to save the entrance station and kiosk, and to prevent the fire from spreading into the gateway community of Foresta. After four days the fire was controlled at 1,840 acres—a figure that 50 years later looks laughably small. It was the threat to Arch Rock, Foresta, and the Crane Creek drainage that had made the fire troubling. A board of review issued 14 recommendations, all intended to improve fire suppression operations. To a later era some seem obvious ("that a qualified fire boss be assigned at the outset of each fire"), and some perplexing (that the NPS change policy "so that cigarettes can be issued to firefighters as part of the normal rations on a fire").[3]

Although wildfire dominated headlines, the future was fire restored. A fully fledged era of fire management for Yosemite, however, took decades to mature. “Progress,” as Jan reflected after his retirement, “came in spite of the agency.” But come it did.[4]

• • •

Bob Barbee thought the “fire thing” the most profitable point of ignition for the reformation of resource management overall at Yosemite. As so often on public lands, fire led. But it remained to transform that intuition from policy, vision, and inherited cultures into a new order of fire. From 1968 to 1971 Biswell and students conducted a suite of prescribed fires in the valley and around Wawona with the assistance of the fire control program. In 1970, coinciding with Barbee’s environmental restoration program, the park conducted its first management burn. Two years later, on May 1, 1972, the park launched its natural fire program. The vision was bold, as befit a flagship national park, but cautious, in keeping with Park Service tradition. It had more checks than balances.

Barbee left Yosemite, reluctantly, in late 1970, passing onward to an extraordinary career that would find him superintendent at Yellowstone during that park’s notorious 1988 “summer of fire,” which in retrospect made the trials at Yosemite seem like a stroll through an outdoor museum. But the Barbee-Biswell duopoly had galvanized enough change that the program had momentum and persisted. Resource Management moved from a desk in the attic of the admin building to a program with a budget, staff, and tasks. There were test fires in the valley, the grove, and the wild. The vacant resource management position was filled. And as Barbee was departing, a Biswell doctoral student was hired as the park’s first research scientist. It proved an inspired choice.

Jan van Wagtendonk did for park science what Barbee did for resource management. He not only channeled Biswell but reconciled fire with wilderness, an issue that had not much clouded the Doc’s determination to get fire back on the ground. He spent his entire career at Yosemite, and remained even after formal retirement, and in its archives reside his personal papers. The finding aid to those papers includes an image, obviously selected by him, of his grinning younger self with a special backpack frame hauling a four-foot-by-eight-foot sheet of heavy plywood into the backcountry. The first instinct is to marvel at his sheer physicality. The second is to ponder how he imagined he could balance such an unwieldy object up and down trails and through forests without fatal stumbles and falls. Yet that is exactly what he and his compadres managed to do.

They did something few had attempted in a systematic way. This was neither let burning nor light burning. No landowner could afford to ignore fires that might bolt beyond their land, and while prescribed burning might be acceptable in a landscaped setting like the valley or grove, a drip torch hardly conformed to the notion of wild lands "untrammeled" by humans. A prescribed natural fire tried to square that circle: it massaged natural ignition with written criteria that specified under what circumstances the fire might continue. Someone had to write those prescriptions. Jan did, as his dissertation under Biswell. Even trickier, those fires did not exist just to reduce fuel or promote regeneration; they existed to sustain "wilderness character." They were part of a suite of mediated interactions between humanity and the wild.

Like many of the early advocates for restoring fire, Jan van Wagtendonk had a varied career with little foreshadowing for what he would actually do, and like many of his contemporaries, he came to fire through suppression. His parents were immigrants from prewar Netherlands; his father, a biochemist, found temporary employment in various American universities until he secured a permanent professorship at Indiana University. Jan was born in 1940 in Palo Alto, California, but grew up in Bloomington, and enjoying the woods, went to Purdue University with a vague notion of majoring in forestry. He got summer jobs fighting fire with the Umpqua Hotshots in Oregon, smokejumped out of the Siskiyou National Forest and Alaska, then transferred to Oregon State University. In 1960 his father sent him an article on fire ecology by Charles Cooper in *Scientific American* that piqued his curiosity—a chance to guide Jan's enthusiasm for firefighting into science. He graduated in 1963. Thanks to the Reserve Officers' Training Corps, which he had begun at Purdue, the next four years he served as an officer with the 101st Airborne, was sent to Monterey for language training, and then shipped to Vietnam, where he became an artillery adviser to the South Vietnamese Army. When his tour ended, he was "selectively retained" by the army and rejected for graduate school at UC–Berkeley.[5]

Any fire story is full of the unexpected and the serendipitous. Certainly Jan's was. He was able to get stationed at Fort Ord, which allowed him to visit the Berkeley campus. By happenstance he met Biswell, who was beginning his fire projects at Whitaker's Forest; Biswell accepted him into his program, and mentored him through his degree; in return, Jan translated the Doc's intuitive understanding of fire into quantifiable prescriptions. ("He knew when to burn and how to burn, but [you couldn't] transfer that to somebody [else]. . . . Because he was an artist.") Those field trials became the basis for his dissertation. Jim Olson, the park fire officer, was willing to experiment, and assisted. In 1972 Jan graduated with a doctorate in wildland resource science, with a specialty in fire ecology. He had studied fire in Yosemite, and the park hired him as a research biologist. But Olson left in November; Don Cross replaced him as fire officer and forced Resource Management to support itself without the use of the suppression program's equipment and capabilities. The program stalled. Jan was diverted to wilderness issues, eventually contriving the wilderness trailhead quota system still in use, and to the preparation of a new master plan for the park.[6]

His role, he determined, was not to decide what the park should do but to provide information and monitoring for what decisions were made. He introduced a geographic information system; continued to consult on the park master plan; and became chronicler, adviser, and guru to the fire program, monitoring operations much as observers did prescribed natural fires. Like other NPS scientists he was corralled into the National Biological Survey in 1993, which was itself soon reconstituted as the Biological Resources Division of the U.S. Geological Survey. While he continued to sit at the same desk as before, the transfer broke his formal conduit to the park. Still, he continued at Yosemite as an informal adviser, and remained something of an oracle even after he retired in 2009.

His influence in the national fire community grew. In 1995 he joined the effort to distill the many agency-specific policies into a common federal fire policy, a project strongly influenced by his Yosemite experience. Beyond that there was so little known of fire ecology or packaged in a form that was accessible to managers—once again he found himself translating intuition into formal scholarship. The founders had been keen-eyed naturalists who understood fire's biological role as Biswell did the practice of burning. That knowledge had to be expanded and translated into usable concepts. In 1997 with Neil Sugihara and Jim Agee he began informal gatherings to discuss California's fire ecology, discussions that subsequently evolved into the Association for Fire Ecology, complete with conferences and a journal (*Fire Ecology*), the first professional society and journal devoted to the topic. In 2006 they, along with others of their cohort, published *Fire in California's Ecosystems*, a majestic survey of California's pyrogeography, revised in 2018, by then featuring 65 contributors. Some 142 years after Yosemite was first set aside and sought to control fires, 116 years after Secretary John Noble had laid down fire bans and sent the cavalry to enforce them, not to mention 300,000 years of human experience using fire, Yosemite had a formal survey of how fire actually worked in and around the park. Unsurprisingly, the book was dedicated to Harold Biswell.[7]

The trek to Illilouette was Jan's conception. Lamentably, when it finally happened, after covid delays, he was terminally ill and present only through the landscape his zest and resolve had helped create.

• • •

The whitish-stone dome is warm, the sun's rays unblinking. Scott Stephens stands under the shade of a Jeffrey pine. The rest of the group fans out in search of various boulders, roots, and granite lumps with some shade. The curve of the exfoliating stone sculpts a rough amphitheater. Behind Scott a larger, counter-amphitheater looms—a scene dappled with burned stumps, shrub fields, and granodiorite. He describes what he and his lab have learned in the Illilouette over the past 20 years.

• • •

The University of California–Berkeley has always had a bond to Yosemite; Scott is its latest face for fire. A Napa native, with degrees from Cal State–Sacramento in electrical and biomedical engineering, he got his doctorate at UC–Berkeley in wildland resource science, with a specialty in fire. He teaches fire science, serves as director of the California Fire Science Consortium, and holds the Henry Vaux Distinguished Professorship in Forest Policy. He has testified to the California Legislature, California's governor, the U.S. Congress, and the White House Office of Science and Technology Policy on fire, climate, and the wildland-urban interface. In 2016 he served on the state's Little Hoover Commission to devise a statewide plan for forest management. He has worked with Senator Diane Feinstein to shape legislation to increase forest resilience in a changing climate. He knows only too well what fire exclusion has meant to California. His research in the Illilouette shows what an alternative future might look like.

Colonization and the creation of the park erased traditional knowledge, not least about fire. The park would have to reconstruct a new regime of understanding from the experiences of the park staff and the field plots of scientists, and both would play out amid deeper cultural perceptions. To most elites and scientists of John Muir's day the scene before us would have appeared as fire-marred if not fire-trashed. To the trekkers it spoke of renewal. The black spikes of burned Jeffrey pine, the green woods bunched like pieces of a biotic quilt, the low ground cover of bear clover tell of good fire restored and a biota rejuvenated. The same scene, different appreciations. But cultural perceptions by themselves lacked protocols for managing; and science alone had no frame for its data points. Its scatter diagrams needed a regression line called a narrative.

The mystery, Scott reiterates, is the Great Disappearance. For a century fire ceased to scar trees throughout the basin. The data recorded in burned boles, in the pyrogeography of biotic patches, in soil and shrubs did not explain why or how this occurred. There is some evidence that a similar process happened more widely, but not widely enough to attribute to climate alone. That would require a major shift in the rhythms of wetting and drying, and the patterns of lightning, and it would have to account for the sudden return of fire in the 1970s. Eliminating the improbable left humans as the likely agent. A study site on the adjacent Stanislaus National Forest had found fire scars 221 times between 1454 and 1912, or a fire every other year. "It is doubted," as park forester Emil Ernst commented, "that Nature could be so regular with fire for such a long period of time."[8]

In truth, the human history of the Illilouette, or for that matter, the park outside the valley and grove was poorly known. No less than for Indigenous societies, the record of backcountry use by herders, hunters, prospectors, homesteaders, loggers, and other newcomers was but sparsely documented, and then almost always to condemn those practices. As the park defined a new purpose for the land, it reconfigured its infrastructure, and then its historiography. The story of Yosemite became one of recreational use, and the tensions between competing claims by concessioners, campers, climbers, and others.[9]

The old order faded away. Trails were repurposed or rerouted, cabins to house the patrons of packing trips replaced those used by cattlemen and shepherds, maps reflected not seasonal browse or game but scenic vistas and climbing routes. Wood signs replaced

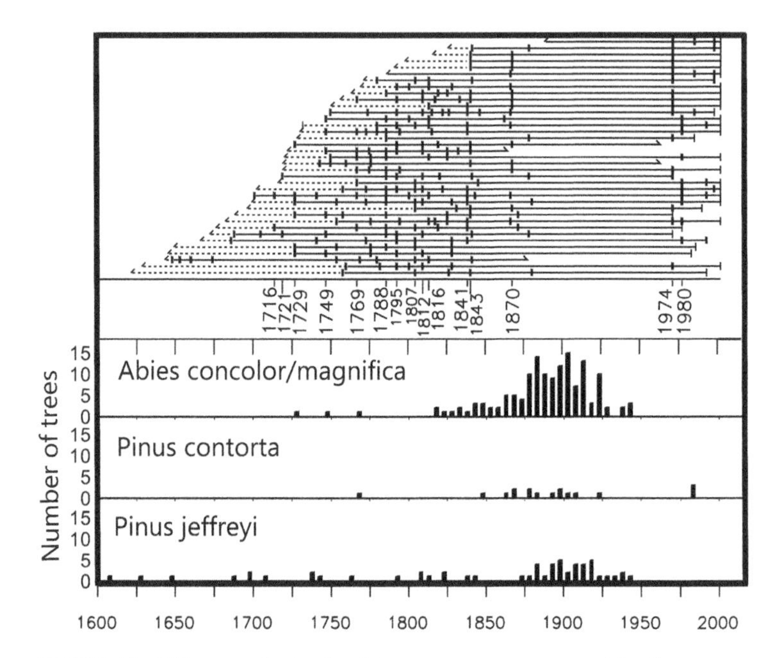

FIGURE 22 Illilouette Basin fire history, as reconstructed by fire scars and stand ages. Each horizontal line chronicles one tree; each tick mark on that line, a fire. Below, for the same years, are the numbers of white fir, limber pine, and Jeffrey pine. Note the Great Disappearance, during which fires leave and the demographics of the three tree species flourish.

blazed trees. The nuances of places that had mattered to Miwok hunters, or later that guided cattlemen to upper-elevation sites that had avoided glaciation and shepherds to glaciated ones, were erased and new stories, like new signage, written over them. Yosemite was a palimpsest of human history, but only the most recent was apparent or chronicled in print and paint. Some prime trails persisted—the logic of the land was too great to ignore. Others were overgrown. As winter recreation arrived, it inscribed further entries into the annals. Eventually, a modern geography of seasonal use overwhelmed the shards and scraps of the past, most of which were recorded in legends, songlines, tradition, oral culture, and in woods, shrublands, and bear clover, all of which could vanish without constant renewal. In its modern avatar Yosemite's history became its history as a national park.

The chronicle of the Illilouette basin is murkier than many. But the Mono Trail is among the most ancient in the park, part of a network that linked Indigenes across the Sierra summit. It later brought summer cattle and sheep that sought out meadows from the west branch of the Illilouette Creek to Merced Pass and Mount Starr King. The era of cavalry administration limited but did not abolish all grazing. Some returned during the hiatus of the Spanish-American War, and during World War I the basin was reopened to grazing to bolster national production. "Even so," a wilderness survey concluded, "use of the area was infrequent." Nor is there much overt sign of Miwok activity.[10]

The major evidence is a massive shift from the meadows that had fed sheep and cattle to dense woods. "The undergrowth around parts of Starr King is so thick and the ground so littered with deadfall that you can sometimes walk a hundred yards at a time without stepping on the ground." The reason is the result "in large part from the removal of an active natural agent, fire." Fire scars cease; forests thicken, meadows thin; backpackers succeed herders, and propane camp stoves replace burning logs.[11]

Then, in accordance with park goals, fires began to return. Their reentry confirmed some intuitions and startled with others. In 1974 the Starr King fire rambled across 3,400 acres in the Illilouette basin. This did not fit the notion of isolated snags smoking like chimneys, or surface fires slinking off into the Tuolumne granodiorite to die. The Starr King fire was nearly twice the size of the Canyon fire. Reintroduced fire would be in the park's face: the flames outlined Half Dome, evening smoke poured over Illilouette Falls into the valley. Fire rose in the league table of park crises.

Oddly, perhaps fortunately, that same year the Waterfalls Canyon fire in Grand Teton National Park, another prescribed natural fire, captured national attention as it crept and swept through July and August and its smoke flooded Jackson Hole before autumn rains extinguished it at 3,700 acres. Park Service Director Gary Everhardt stood firmly with the policy and the parks. Grand Teton took the heat as the national media scrambled to unpack an era in which authorities cheered fire rather than quashed it.

• • •

The program pushed on. The Protection Division updated its plan for wildfire, replacing the phrase "takes precedence" with "highest priority," and allowed that prescribed fire "may be employed as a substitute for natural fire," while Resources Management prepared the plan for such burns. Bob Barbee completed his *Environmental Restoration Program for Yosemite National Park: 1970,* which proposed extensive work in the grove and valley, with some prescribed fires moving from those anchor points to mixed-conifer forests around Wawona (aided by "professional advice and prescriptions" from Dr. Harold Biswell and Harry Schimke, a Forest Service burning expert). In 1972 the plan allowed for naturally ignited fires in designated zones to roam, modeled on the experiences accrued at Sequoia-Kings; among them was the Illilouette basin. Yosemite fire had passed through a phase change. The Starr King fire was its coming-out party, complete with fireworks.[12]

At Everglades, at Sequoia-Kings Canyon, at Yellowstone, at Wind Cave and Saguaro, and at Yosemite—the early Park Service adopters crafted a tripartite strategy. The built environment required urban-style fire protection, and where communities crowded the park border and landowners, public or private, demanded suppression, the boundary became a full-control fuelbreak. Remote areas could be considered for natural fire restoration, subject to various conditions. The remainder of the park was available for prescribed fire either as a catalyst to reinstating natural fire or for fuel reduction to protect buildings, campgrounds, critical assets like sequoia groves, or boundaries.

At Yosemite, however, the change in strategy was not met with an equal change in organization, so that one plan existed for fire control under the auspices of the Protection Division, and another existed for fire restoration, under the new Resources Management Division. By reducing fuels, a prescribed fire program could serve both organizations. By restoring a natural process, however, a natural fire program could provoke rivalries over when and where such fires could be allowed, and who should decide.

A stand-alone prescribed fire management plan arrived in 1976. Its purpose was clear: prescribed fire existed "to re-establish a pristine condition in which natural fires will be allowed to burn throughout all natural areas of the Park." Some sites like the valley and grove would likely require prescribed burning in perpetuity, but the paramount goal elsewhere was to allow or where necessary to recreate presettlement conditions that could perpetuate themselves indefinitely. The program would first target the valley, the grove, and virgin sugar pine forests. Eventually guidelines would extend to incorporate meadows, chaparral, and mixed-conifer forests—"by far the most extensive and diversely vegetated zone." The more elevated the landscape, the more the plan granted fire freedom to behave untrammeled; a zone of "conditional fire management," of mixed responses because they felt most acutely the consequences of fire's suppression, occupied mid-level elevations. To oversee the program the plan provided for a Prescribed Fire Committee comprised of the superintendent, and one member each from the Division of Resources Management, Division of Visitor Protection, Division of Interpretation, and the Fire Management Office.[13]

They were heady times. Despite bureaucratic frictions and pushback from fire officers for whom all-out fire suppression had defined not only their careers but often their sense of themselves, Congress and the public shared an ardor for the wild, and the national fire revolution was approaching its high-water mark. Upgraded plans for prescribed fire and resource management appeared in 1977. Another update in 1978 revised the boundaries of fire management units and expanded the realm of prescribed natural fire; the suppression-only zone shrank to less than 20 percent of the park. In 1978 Stephen Botti and Tom Nichols, resource management specialists for Yosemite and Sequoia-Kings Canyon, respectively, summarized the founding decade of prescribed natural fire. Some 84 percent of Yosemite and 94 percent of Sequoia-Kings were managing some variety of prescribed natural fire—Yosemite's "conditional zone" allowed lightning fires to burn after September 1, when the fire season had historically begun winding down. That same year the Forest Service scrapped the 10 a.m. policy in favor of a major overhaul of its fire programs.[14]

Yosemite's PNFs were mostly small: 67 percent were under 0.1 ha; and 4 percent greater than 120 ha. Most burned for a few days; others longer, up to 108 days; Starr King, for 72. In 1978 the park had seven prescribed natural fires, each over 80 ha, burning simultaneously. One of the happy revelations of the decadal experiment was that "large prescribed natural fires have continued to burn under what are normally considered severe fire weather conditions without requiring any suppression or containment action." Similarly, "no prescribed natural fire has ever required full suppression"; only four had prompted any efforts at containment. "Further expansion of prescribed natural fire unit boundaries remains a distinct possibility."[15]

That was the good news. The bad news boiled out of Rocky Mountain National Park in 1978 as the Ouzel fire, begun as a PNF that hovered around the tree line, bolted eastward under the impress of chinook winds and threatened the town of Allenspark (the park was even cited by Boulder County for air quality violations). A national handbook for fire planning (NPS-18) resulted. It little affected Yosemite since Yosemite's experience was the basis for much of the national standard. The real issue was the tense relationship between the rival visions that prevailed between Protection, which wanted to eliminate bad fires, and Resources Management, which sought to reinstate good ones.

made," and that the park need not prepare an environmental impact statement (EIS). A policy that promoted natural fires in wilderness did not need an EIS (what needed justification was a policy that suppressed such fires). The environmental assessment submitted for the 1977 resources management plan still sufficed. The term "prescribed natural fire," like dissidents in Argentina's dirty war, was disappeared.[18]

It had not seemed to those on the front lines that conditions were benign—not those in nature, not those in politics. Compared with the recent past, these were scary burns that pushed the envelope of administrative control. But in retrospect a fire like Starr King that burned 3,400 acres seems like a minnow amid a rampaging pod of orcas. By the mid-1980s the West was shuffling into a mega-drought, the worst in 1,200 years; the reckoning of a century of disrupted fire regimes was at California's doorstep; urban sprawl splashed across formerly rural landscapes. The old controversies and brouhahas over where to place this line or that on a map of fire management zones looked like arguing over seat cushions during a train wreck. Humanity's fire habits had unhinged the atmosphere, and the unmoored climate was acting as a performance enhancer on an intrinsically fire-prone landscape now primed for upheaval.

It was not just the public that had to be reeducated. The park had to accommodate its own traditions and internal divisions, and adapt to a great unknown as California spun into historic, perhaps epic, realms of drought and flammability. Each backcountry fire could be a contest over which strategy should prevail. Regardless of word-smithed guidelines, suppression remained a default choice: no one was chastised for putting a fire out, but letting one become a public nuisance or worse, escaping, could end a career. More and more criteria entered the checklists for prescribed fire, any one of which could prevent ignition. Each new complication that suppression encountered became, by contrast, an argument for more funding. There was no checklist by which to add prescribed fires as there was to subtract them.

The 1979 *Natural, Conditional, and Prescribed Fire Managem* *Plan* consolidated the evidence accumulated since 1970, sharpe prescriptions, and tried to shift the burden of justification to press from those seeking to restore fires to those who instinct raced to extinguish them. "The fire management committe conduct a critique within one month of any suppression or con ment action directed against a natural, conditional, or presc fire; and the Chief of Resources Management will, subsequ submit a written report to the Superintendent analyzing and ating management actions." In principle, natural fires replace pression as the default setting.[16]

The newly established director of fire management for th David Butts, thought Yosemite's plan "excellent," "innovati "one that will be of interest to other agencies." But he was "cc how "it can successfully be administered as two apparently programs" because a "fire management program, suppres use, is indivisible from the natural resources it affects." He re that Yosemite was not unique in its struggles, and was conf it would "continue to provide leadership" in its fire progra dendum to further clarify responsibilities followed in 198

A management rule of thumb holds that most agency zations are, at base, an effort to get or to get rid of son schism at Yosemite was not solved by plans or protoco solved when the critical figures left. In 1989–90 the tw finally merged. By then Yellowstone's Big Blowout had Park Service through a political firestorm. Norm Chri enlisted again to chair a committee to review policy. I as had the Sequoia-Kings Canyon committee two year policy was sound but execution might need attention. (reviewed the season; even the Government Account got into the act. Yosemite modified its zones for pre ral fire but otherwise determined "that the Fire Man is not a major federal action which significantly affe of the human environment," that "no significant im

The fire crisis shifted from reinstating snag fires in the wild and missing ones in the valley and grove to feral flames that posed real threats to life and property like the 1990 A-Rock and Steamboat fires in Yosemite, and the 1994 South Canyon fire that burned over a crew in Colorado amid the country's first billion-dollar suppression bill. The national response, in 1995, was the promulgation of a common federal policy for wildland fire management, and subsequently the 2000 National Fire Plan. Locally, the 2003 Kibbie Complex jointly handled by the park and the Stanislaus National Forest gave extra impetus to a major overhaul.

The National Fire Plan posed a devil's bargain: it dumped bundles of money into fire but toward goals not readily aligned with those of Yosemite. Its announced purposes were to mitigate the excessive fuel buildup that had resulted from fire exclusion, protect vulnerable communities, and reduce suppression costs. It also shunted fire funds that had been under the control of the NPS to a Department of the Interior directorate—the same shift that had occurred in research. The Park Service had to justify its requests in competition with larger agencies, notably the BLM. And it had to reconcile fire protection with fire restoration, this time not at the level of individual parks. The money was welcome. The loss of say over how it might be spent was not. Still, new staff in resource management were hired and funds were funneled into prescribed fire projects.[19]

Over three decades of experience reached a climax when, in March 2004, the park submitted its *Final Yosemite Fire Management Plan*, wrapped in a full-bore environmental impact statement. The recommended course of action became, a year later, Yosemite's operational fire plan. The 1964 fire control plan, considered a model for its day, had run to 39 pages, which included tool inventories of fire caches, a phone directory, and a roster of fire qualifications for park personnel. The 2004 plan had 756 pages of text and perhaps a hundred figures and maps. It probably contained as many words as all the research published in American wildland fire science before the 1968 Green Book.

More specifically, the plan sought to reduce the threat of wildfire to public safety, to built environments, and to cultural resources, and to otherwise "return the influence of natural fire to park ecosystems so that they are restored to, and maintained in, as natural a condition as possible." Audaciously, the plan proposed "to reduce risk to park wildland urban interface communities within six to eight years, and to restore park ecosystems within 15 to 20 years." Mitigation around communities would involve mechanical treatments; the remainder would rely on "prescribed and wildland fire." The annual burned area for the park was estimated at 16,000 acres a year, and on average that is what the plan sought to restore.

The plan was a triumph of the fire revolution and of the generation that had come of age with Harold Biswell and the Leopold Report. Tom Nichols, the park's fire management officer and the plan's principal author, had trained in fire ecology, not forestry; had learned fire through prescribed burning, not suppression; had come through Park Service ranks in resource management, not rangering. He read the Leopold Report as reaffirming the founding directives of the agency, to leave the scene "unimpaired" for future generations. The National Fire Plan assumed that getting fuels right would get the ecology right. Yosemite's fire plan assumed that getting fire's ecology right would take care of the fuels.[20]

The plan was revised in 2008, with the intention of amending it annually. The next year a prescribed fire set in Big Meadow to help protect Foresta flung spots beyond perimeter lines and bolted out of control 55 minutes after ignition, skimming from snag to snag (themselves scars from the A-Rock fire) and overrunning the Big Oak Flat Road. An 89-acre subunit blew into 7,425 burned acres; a fire set to protect the Crane Creek drainage instead blew through it; a community uneasy about the park's ability to shield it from wildfire had its fears seemingly confirmed. This time the prescribed fire program at Yosemite was not compromised (or distracted) by escaped natural fires in Rocky Mountain or Yellowstone but by a prescribed fire ignited by its own hand. The Big Meadow burn prompted a review, followed by one for the park's fire program overall.[21]

FIGURE 23 Fire lighting falters, as the Big Meadow prescribed fire escapes in 2009. Looking down at the scene 12 years later.

The final report concluded that the burn plan had proceeded under policy but was inadequate for the actual conditions present, and that the park's prescribed fire plan had been followed but that its prescriptions were inadequate (they had failed to account for the snags). Action items were generated; the park was allowed to continue with its almanac of burns. Then the devil's bargain demanded its due: California blew up in 2003, 2007, and 2008, and suppression costs went ballistic. No quantitative evidence existed that a broadly defined "fuels" program was working. In 2009 the Office of Management and Budget insisted that 90 percent of the fuels funds made available through the NFP go directly to treat communities at risk. Yosemite faced a double blow. The Big Meadow escape increased the complexity of prescribed burning while the loss of fuels money for ecological restoration reduced the park's ability to do that burning. That same year, however, new guidelines for interpreting federal fire policy improved opportunities to work with—effectively, manage—wildfires. It seemed possible that herding wildfires might accomplish what prescribed fires had not.[22]

A 2011 Fire Program Review lauded Yosemite as "a leader in wildland fire management," able to operate successfully "in a challenging and political environment." But the flush of fuels funding granted under the National Fire Plan had, like the gold placers of the forty-niners, played out, much as had the emergency presuppression funds the Park Service had relied on previously. The flush times had been welcome, but they had allowed the NPS-dedicated budget for fire to wither, and when the NFP funds shriveled there was no agency substitute. The issue was service-wide and department-wide—DOI's Office of Wildland Fire controlled the big money. Yosemite's fire program would have to "rebalance," which meant do more with less. The review team expressed its confidence that Yosemite would "continue to meet these and any new challenges."[23]

An updated plan emerged. A new fire management officer arrived, who soon found much to confront in the "behavior" of parts of her staff, some of whom were already under investigation for drug use, with the result that many left and others were told to transfer (an episode long referred to as "the Purge"). Funding continued to lag. Nationally, the escaped Cerro Grande fire, and locally the Big Meadow burn, still festered. The Great Drought gripped the Sierra, adding unanticipated fuel as hillsides of trees died from beetles, aridity, and eventually fire. Much of the fireshed lay outside conventional fire behavior and prescriptions. With too many unknowns and too few anchor points, the program shifted from hotlining to holding. Some treatments progressed around gateway communities like Hodgdon, Yosemite West, and Wawona, but emphasis shifted away from deliberate burning, for which funding had been lost, to managing natural fires in the backcountry, which could be finessed under a confine-and-contain suppression strategy. Overall, the program faltered.[24]

The 50 lightning fires that on average Yosemite experienced each year continued, and more of them became large. Even in 1975, 1978, 1980, and 1981 worrisome burns had splashed across the Illilouette; but now their enhanced social context made large-area, long-smoldering fires more troublesome. Always there were competing concerns. After burning a month, and smothering the valley with smoke, the Buena Vista fire was suppressed. To potential threats to developments, the 1987 August Complex added the hazards of smoke and was extinguished, with the first of California's fire "sieges" as a backdrop. In 1988 the Walker fire, along the Tioga Road, burned for a month until worsening fire conditions caused it to be contained at 3,450 acres and subsequently featured on NBC's *Today*. Meanwhile, Yellowstone's Big Blowout raised enough political clamor that the Park Service banned further prescribed natural fires until parks rewrote and resubmitted their fire plans for approval. The ban affected Yosemite for only a year, but its shadow passed over the national agency like a smoke pall. What Bob Barbee had kindled in Yosemite had, it seemed to many, exploded in Yellowstone.[25]

By the 1990s the Illilouette had received enough fires that new burns began to enter the scars of old ones, and in cases to layer and knead into one another, and appeared to confirm the hopes of the program's founders that restoring fire would restore fire's capacity to self-regulate. Nature would reclaim control and allow the fire program to shift from active management to a more custodial role. But less benign burns crowded that story into sidebars.

FIGURE 24 Fire fighting fails as suppression seemingly turns on itself: Miguel Meadows guard station destroyed during the Rim fire.

The Merced River, in particular, was becoming a flue for fire. In 1990—Yosemite's centennial as a national park—lightning kindled 27 fires, two of which near the park's west entrance blew past initial attack. The A-Rock fire raced toward El Portal, Crane Flat, Big Meadow, Foresta, where it burned houses, and forced Yosemite Valley to evacuate. For the first time in its history, the park closed. The Steamboat fire kindled in the Merced River Canyon and threatened Yosemite West and Badger Pass. The Illilouette got its version with the 1991 Ill fire and 1994 Horizon fire, both of which were eventually suppressed after saturating the valley with smoke for weeks before finally escaping their confinement zones (at 3,814 and 3,860 acres, respectively). In 1996 the Ackerson Complex absorbed 13 lightning fires along the boundary with the Stanislaus National For-

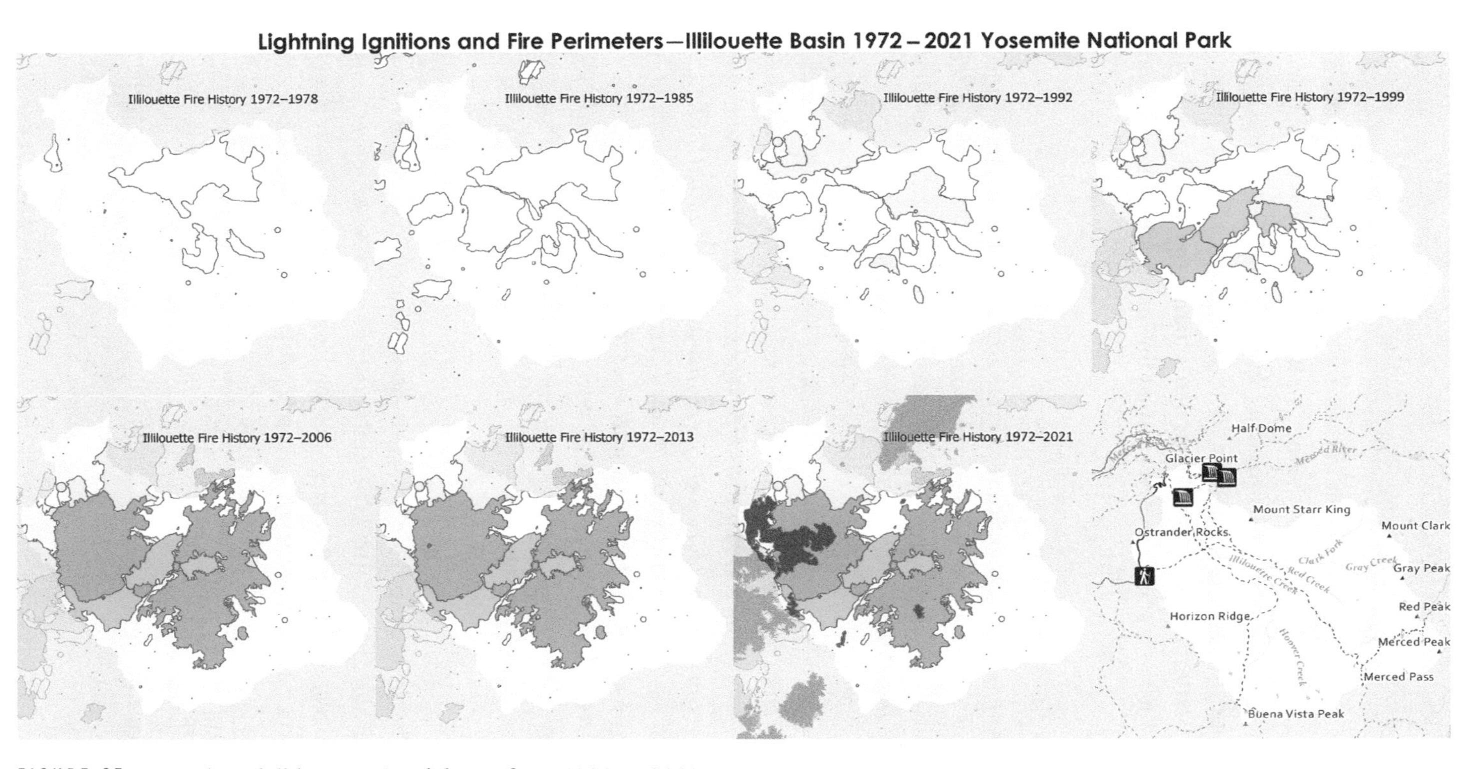

FIGURE 25 Fire atlas of Illilouette Creek basin from 1930 to 2021.

est, burning 47,000 acres in the park and 11,000 in the forest. In 1999, while Secretary of the Interior Bruce Babbitt was announcing the advent of a national fire crisis, Yosemite managed 12,445 acres from multiple fires, all under a doctrine of *wildland fire use*, the successor term to prescribed natural fire, until it too was disappeared as yet more guidelines were issued for interpreting the common federal wildland fire policy.[26]

• • •

It's midmorning when Scott leads the group to another dome farther east. This one offers a sweeping panorama of the ramparts of the eastern Illilouette, from Mount Starr King past Merced Peak and around Mounts Clark, Gray, and Red.

Scott explains what to expect as we leave the trail for a broken-country hike to a meadow beneath Starr King. The result, he admits, can look messy to the untrained eye. "It isn't always clean," he says of the forest, "and it's not always nice." The scene has traded traditional expectations about landscape for resilience. It alerts us that the future will need new aesthetics, an updated descriptive language for landscape, and an appropriate narrative as well as refashioned science.[27]

As we leave the rise, a veil of smoke appears along the south, the outlier haze from Sequoia-Kings Canyon as the Colony and Paradise fires prepared to merge.

• • •

With reintroduced fire the forest monoculture had fractured into a kaleidoscope of shrubs, meadows, and patchy woods of Jeffrey pine, lodgepole pine, and scraps of aspen. Forests over which 20 years earlier it was possible to walk a hundred yards without touching the ground were opened; forest cover in the basin overall shrank from 82 percent to 62 percent. Wet meadows had recovered. Snags and charred trunks abounded. Basin hydrology saw a 30 percent increase in summer soil moisture, adding the equivalent of two inches of water.[28]

FIGURE 26 An aesthetic for the Pyrocene: what a healthy, resilient forest in the Illilouette basin looks like today. In addition to novel science and policies, Americans will have to cultivate means to appreciate their evolving pyrogeography.

The scene was unusual for contemporary California, but not for the Sierra de San Pedro Martír in Baja California, where Scott had collaborated with Richard Minnich from the University of California–Riverside. Minnich had for years observed that north of the border, wildfires were few but immense; and south of the border, many but small. Fire suppression ruled in Alta California; historic conditions that favored lots of fires mostly persisted in Baja. Missions in the 1790s had introduced disease and livestock, and in places broke the old rhythms, but fire suppression was only attempted late in the twentieth century, and much of the mountains still approximated presettlement conditions. After 50 years of partially restored fire, the Illilouette now displayed fuel loads and basal areas comparable to the Sierra de San Pedro Martír. Both seemed to be riding through climate change with more ecological integrity than their peer landscapes, whether private or public. The Sierra Nevada needed a future that looked more like the Sierra de San Pedro Martír and less like the Transverse Range that framed Los Angeles.[29]

The theory had always been that as fires moved from restoration to maintenance mode, they would check future fires, or at least calm their behavior, that the firescape would become self-regulating, with old patches delimiting future ones. The Berkeley research both confirmed and complicated that presumption. The chronicle recorded in fire-scarred trees had seemed to point, on average, to a presettlement periodicity of roughly eight years; what the Illilouette research discovered was that old burns up to eight years old could effectively halt new burns, but that burn footprints nine years or older did not. Old burns influenced severity and intensity, but not spread; after eight years new fires could burn over the former scars readily. Multiple overburns made no difference.[30]

In brief, research confirmed theory: old burns did affect future burns. But it was not clear that theory meant what it did when the program started. In 1972 it was assumed that lightning fires would create a jigsaw puzzle of mutating pieces, each affecting its neighbors, amid a more-or-less-stable regime. But the Illilouette had experienced significantly fewer fires than historically. From 1972 to 2005 some 28 percent of all starts had been extinguished, most of the larger fires were deliberately halted before rains ended the fire season, and all human ignitions had been suppressed. Already in 2021 two lightning fires had been put out before they could spread.[31]

Gradually, as California's climate assumed more extreme forms, as serial conflagrations slammed into the countryside year after year, a sense of dread grew, like a cold pit in the stomach, that the old order might not return. The premises of the fire revolution that a reformed policy and a once-off correction in the field would rebalance landscapes that could then self-maintain in the ways they presumably had in presettlement times shimmered like a mirage, always just beyond reach. In Northern California the Carr fire burned 220,000 acres and 1,000 homes in 2018, then the Zogg fire reburned 56,000 of those acres and 200 structures two years later. Such fires were not midcourse corrections, but catalysts for a new trajectory in which fire begets fire. It seems fire management could not be done once, then tinkered with, but would demand active engagement in perpetuity.

Restoration had always been a vexing concept, intrinsically prone to irony. At Yosemite it meant one thing in the valley and grove—the maintenance of a historic moment as biotic museum pieces. In the wild it had come to mean restoring fundamental processes, of which fire was the most broad-spectrum catalyst. As a goal, resilience was replacing restoration. It mattered less that fire officers address an out-of-whack firescape that seemed destined to depart ever further from historic norms than seek to preserve as many ecological goods, services, and processes as possible.

That is what happened: the Illilouette was more resilient than most Sierra landscapes, though perhaps not enough to match historic dimensions or to survive a fast-morphing climate that was rewriting the rules of fire behavior. The Illilouette experiment had begun in what a later generation of fire folk regarded as a benevolent era and then coevolved with changing conditions. What mattered was less the specific character of the Illilouette biota than its ability to absorb continued insults. However troublesome fire had become over the course of the 50-year experiment, the challenges were almost certain to get worse in the future.

• • •

The trekkers pick their way over the creek, across fallen trunks, and through dense meadow grass. At an open patch someone finds an arrowhead. Eventually the group halts along a copse of aspen to eat lunch and chew over a scene in which, while California struggles against megafires and smoke palls, a basin has reduced the likelihood of blowups, improved biodiversity, and boosted water retention.

On their return the southern horizon of smoke has thickened, ready to spill over the Buena Vista Crest like fog.

• • •

The Illilouette basin had been chosen for attention not only for its granitic borders, but because it was not along the park boundary, not an iconic landscape, and not subject to heavy backcountry use. It was not a destination landscape. Which is to say, it did not require shared jurisdiction and negotiations with neighbors, would not attract devotees alarmed over changes to a deep-memoried landscape, and would not arouse outrage when parts had to be closed because of fires. Though easily visible from Glacier Point, it was largely invisible to Yosemite's constituents. It was known through smoke carried to the valley with down-drainage flows each evening, and that occasionally burst into public attention when its plumes towered above or its flames backlit Half Dome.

The Illilouette fire project stood to the wild as the Mariposa Grove did to the park's sequoias. When the California Wilderness Act placed 95 percent of Yosemite into formal wilderness, the Illilouette was no longer generic backcountry, and it acquired protocols beyond what the park had established in the aftermath of the Leopold Report. The Green Book set goals, but did not mandate methods, and it left considerable discretion with park authorities. The Wilderness Act subjected the park's decisions to standards prescribed by law and subject to public review in court. It made the Illilouette first a wilderness, and second an arena for fire management. Fire officers might wish that relationship was reversed, might argue that active fire management was a primary means to ensure wilderness conditions, but the law said otherwise. Whatever decisions would be made regarding fire would have to conform to the "wilderness character" of the Illilouette.

That significance was much on the minds of Dan Buckley and Kelly Singer, Yosemite's chief and deputy chief fire management officers. Two months earlier, on July 4, lightning had kindled a fire in the Mokelumne Wilderness within the Humboldt-Toiyabe National Forest along the California-Nevada border. A single snag, some ground fire, contained within a granite basin, isolated in a legal wilderness—the Tamarack fire wasn't much of a burn, and the Forest Service chose to monitor it rather than send in crews. There were at the time onerous suppression projects underway elsewhere in the northern Sierra—the Dixie fire had even crossed the summit of the Sierra Nevada twice. Then the Tamarack fire blew up, boiling over 68,637 acres by July 28, overrunning small hamlets, putting Alpine County into disaster status, and massing into a political firestorm that helped push the Forest Service's Chief Randy Moore to prohibit any prescribed fire or natural fire use for the remainder of the season.[32]

It was a draconian step, probably necessary, but it came without any indication if and how the agency proposed to make up those lost acres of good fire. The episode demonstrated, yet again, that it was acceptable to lose an initial attack or repeatedly fail to control a wild fire, yet not to lose a prescribed or monitored fire under any conditions. The reality is, all strategies of fire management suffer failures. The rate of escapes for suppression is 2–3 percent; for prescribed fires, half of that or less (the Forest Service claims less than 1 percent, the Park Service 1.5 percent). But prescribed burns come with liabilities that suppression fires don't confront. In a deeply divided American politics looking for sparks to kindle memes, every lapse is primed for a blowup. By mid-July 2022 no escaped suppression fires were criticized, but two prescribed fires in northern New Mexico that merged to become the Hermit's Peak fire blazed across national news and resulted in a 90-day, countrywide moratorium on prescribed burns while the Forest Service reviewed policy—effectively shuttering another season for restoring fire by controlled means.[33]

Still, wilderness, rightly, has its claims. They, too, are political and they have legal standing.

• • •

The trekkers halt by a shaded patch of forest, and position themselves, as though around a campfire, while Mark Fincher explains what wilderness means. Much as fire synthesizes its surroundings, so wilderness integrates the sensibilities of its time. The case for restoring fire seems self-evident; surely, a natural process allowed to proceed in a natural area without intervention meets any coherent definition of "untrammeled." A violation of wilderness character would result from removing such a fire, not tolerating it. But one variable chasing another variable does not yield a constant.

• • •

Mark has spent his career—30 years—engaging Yosemite's wilderness. For him this is not a job: it's a vocation. He has not only read the Wilderness Act but the congressional hearings that led to it and expanded it, along with the court rulings that have sharpened its legal status and the scholarly commentaries that have sometimes illuminated and often obscured those texts. He knows wilderness as an idea, as a body of legislation and court law, and as a practical reality. Every wilderness job in the park he has done at some time, and now as designated wilderness specialist, he is the resident authority on what is or is not appropriate in wilderness, the go-to guy for deciding about ecological interventions, recreational use management, commercial services, even scientific research, in wilderness, and of course what fire practices might or might not be appropriate. If it is John Muir's words that partisans of Yosemite's wilderness read, it is Mark Fincher's acts that have given those texts substance.

Interpretation, he elaborates, hinges on that liminal word "untrammeled" (reputedly taken from a passage Howard Zahniser had read about the ocean). But the essence is a notion of "wilderness character." What enhances wilderness character is acceptable; what degrades it is not. Wilderness is the place where humans choose to do as little as possible. They don't instinctively intervene; they exercise restraint, not flex their technological prowess. Perhaps Howard Zahniser's choice of untrammeled was inspired because the motives and cultural itching of a society that prides itself on "can-do" is as bottomless and uncontained as the sea.

That means Mark's task is primarily one of saying no. Every group, every American, believes their cause is exceptional and the demand they make on wilderness is just. That, of course, is a formula for decay. Aldo Leopold famously noted that it was the burden of an ecologist to see wounded landscapes. It is the particular burden of Mark Fincher to preside over wounded wilderness. He knows that preserving the wild demands constant negotiation, he appreciates that every concession spends not interest but capital, he accepts that he must hold the wilderness ideal as a north star while navigating the incessant winds and waves that make wilderness travel unique, eccentric, endless, and necessary. He has joined the trek because fire is one of those voices clamoring for exceptions.

From 1930, when records began, until 1972, when the natural fire program commenced, the Illilouette experienced 100 fires that burned a total of 26.7 acres. From 1972 to 2000 it had 118 fires that burned 7,489 acres. Yosemite's fire atlas shows large swathes of the park burned in recent decades, but from big fires like the Rim and Ferguson that blasted into the park across its western border. The Illilouette shows a different pyrogeography: smaller fires, with an occasional large one, that burn into and over one another. Still, of the Illilouette's 139 lightning-kindled starts, 32 percent of lightning-caused fires (28 percent of all fires) have been suppressed. In 2021 lightning ignited half a dozen fires in the basin or along the Glacier Point Road. The King and Clark fires were extinguished at 10 acres each, and the Hoover fire at 59 acres, along with a suite of small ignitions, including two suppressed a few days before the trek.[34]

Smoke is the principal reason—that and demands for fire suppression resources outside the park that either the park must commit or that are unavailable should the park have other starts that threaten major assets and need to be suppressed as quickly as possible. In Muir's day, the crisis was the O'Shaughnessy Dam flooding Hetch Hetchy Valley with water. In Mark's day, it's smoke flooding Yosemite Valley. In 2013, as it burned into the park from the west, the Rim fire poured so much smoke into the valley's granitic box that the air quality index blew past 500, which put it beyond "hazardous." A park that had only closed twice in its history, once for fire and once for flood, now had to close for smoke.

This summer the fire program has nurtured two largish fires along the Tioga Road. The County Line and Lukens fires are contained to the north by forest fractured with patches of granite and on the south by the scars of the White Wolf and Blue Jay fires from 2020. It is a textbook case of using past fires and terrain to encourage more good fire, but the cost was the decision to extinguish those new starts in the Illilouette.

Of the 16,000 acres roughly estimated to be fire's historic burn rate annually, most have come, thanks to the wizardry of averages, from uncontrollable megafires like the 2013 Rim fire that crashed into the park. The acres from prescribed fire are small, though not trivial—148 acres burned in the valley, for example—serve multiple purposes. But prescribed burning is shrinking, mostly confined to valley and grove or to piles from mechanical thinning projects. The future resides in managing wild fire. Fire officers want to boost wilderness burning to help maintain ecological integrity and to avoid what Harold Biswell was prone to call a "holocaust" fire. Wilderness rangers want to stay those fidgety hands to help maintain wilderness integrity. Big burns are likely in the future, whatever people do or don't do. The wild can tell us how nature will accommodate those outbreaks in the absence of our interventions.

The negotiations are friendly, though they can be strained. The trek to Illilouette was designed to allow both parties to discuss their

FIGURE 27 Smoke from fire in the Illilouette pours—inevitably, inexorably—into Yosemite Valley.

wishes away from offices, beyond ideologies and personal biases, on the ground where fire and wild nonchalantly perform their peculiar choreographies.

• • •

There are endless reasons to stop a fire, none subject to public criticism. There is only one to allow it to persist. However natural its ignition source, however squarely embedded in legal wilderness, liability resides in the person or agency who decides not to extinguish it. When Hank DeBruin sought to reform Forest Service fire policy in the early 1970s, Chief Forester John McGuire said he would agree, if DeBruin could do it in five years and not suffer any fatalities. But "if you lose one person in this process, you're done." DeBruin thought it "a fair risk." Lots of others have thought otherwise.[35]

During the 1994 season, the Horizon fire, a troubling PNF in the Illilouette, sparked discussions between Jim Sullivan, the park's fuel technician, and Dave Allen, the fire officer responsible for PNFs. His thoughts were, Sullivan explained, all "gut feelings," but in retrospect they may be all the more realistic for that reason. He foresaw three alternatives. One, blackline the greatest allowable perimeter and let the fire expand within that domain. Two, go indirect, as the park was doing. Or three, go direct "and put it to bed." Why suppress? It was a demanding fire season throughout the West. Weather and fuel conditions were worse than normal. The NPS was struggling with a troubled PNF elsewhere and didn't need more unwanted political attention. National suppression resources were fatigued; hotshot crews in the Southwest had not had days off for months. The historic fire season for Yosemite was approaching. Recent fatalities were a "political hotspot." "Don't set yourself up." Bottom line: "You would gain more in credibility than you would lose in acres accomplished if you weighed all these factors and used them to justify putting it to sleep." Yes, the area needs to be restored to a more natural state. But not this place at this time. If you succeed, the agency gets the credit. If you fail, the penalty falls on you. The reasoning was impeccable, and in various re-creations it has remained unchanged for decades.[36]

The checklists of go/no-go decision points swell with each year, with every new social value identified, with every glitch and stumble in fire practice, and with every personnel transfer into and out of the park. Fire management must cope with the Clean Water and Clean Air Acts, with the Endangered Species Act, with cultural resources, with Wild and Scenic Rivers, with visitor safety, with the regional fire danger and available resources, with personalities among cooperators, a punishing drought from 2012 to 2016 that colluded with beetles and killed 160 million California trees—and on and on. Gateway communities and concessioners. Invasive plants. Callouts under the California Fire Master Plan that can divert crews from fuel and prescribed fire projects. Wretched air in the San Joaquin

Valley. Disturbances that are compounding with one another. Uncertain and unequal funding among aspects of an integrated fire management program. Fire response varies according to whether the flames are at the boundary, the Pacific Crest Trail, the Tioga Road, Hetch Hetchy Reservoir, Wawona, El Portal, or the Yosemite Valley. To stabilize some of those constraints was a purpose of the 2004 environmental impact statement.

The upshot is that complications increase along with urgency. The need for good fire, and the likelihood for bad fire, is matched by the number of legal, bureaucratic, social, and political hoops that must be jumped through. There is always a reason to delay, to defer for another year, to keep flames out of this particular site this particular season; yet no equivalent mechanism exists for recapturing that lost burn. What wilderness partisans see as death by a thousand cuts, a scenario for slow, then quickening ruin, applies equally to fire officers, who see each delay as compounding interest toward disaster. Fire didn't seem easy in the 1970s, but circumstances are far worse today, and they will be even more hostile 50 years hence. What seemed like edge-of-the-envelope risks at the time can seem like missed opportunities decades later.

Options do exist. The Sierra Nevada parks developed the concept of minimum impact suppression tactics—ways to engage fire without inflicting more damage by fighting the fire than the fire itself might cause. Well-placed shovels instead of bulldozers, waterbars in firelines instead of erosional gouges, and constructive use of terrain are all means to lessen the intrusion, which can be especially brutal where tracked vehicles operate. It should be possible to create a parallel suite of practices for prescribed fire that could pass wilderness review.[37]

Of special interest is the concept of relighting or catch-and-release. The two fires that kindled in the basin were suppressed because they came early in the season, would smoke in the valley perhaps for months, and occurred when California was scrounging for firefighting resources. Had they come later in the season, the park could have accepted them. Why not *re*kindle them later? Restart them where they ended and let them propagate outside the constraints that made them unacceptable. Here was an intervention, not a natural act, but perhaps a compromise that would create space for fire management without violating wilderness character. Dan Buckley wants to push the idea into field tests. Mark Fincher is willing to talk. Were he present Jan might note, as he did during an oral history, "You know, the Park Service changes its policy first and then tries to figure out how to deal with it, rather than the other way around." Neither wilderness nor fire is logical. Probably they never will be.[38]

But fire does not limit its impacts to wilderness. In 2014, near a meadow on the north side of Mount Starr King, just across the divide from the Illilouette watershed, lightning started a fire that simmered for three weeks until latch and key—wind and fuel—clicked into place and blew the fire over nearly 5,000 acres in Little Yosemite Valley. Before rains ended the Meadow fire, its smoke filled the valleys, its flames lit the evening sky, and helicopters were evacuating 85 hikers from the summit of Half Dome.

Meanwhile, as the trekkers continue their discussions on bog and granite, the fires at Sequoia-Kings have merged into the KNP complex. Ash Mountain is preparing for evacuation. The flames will likely reach Giant Forest.

• • •

An expectation for the trek was that the mixed group would gather around an evening campfire and discuss how our separate understandings could be joined over the shared fire. In nature fire integrates its surroundings, so in a social setting it brings people together. Sitting around a fire is as elemental an experience as a human can have: we're drawn to the flames. They mesmerize, they mellow. Fire is what we do that, presently, no other creature can. It's our ecological signature. Demanding that a human not use fire is like insisting that a wolf not hunt or an elk browse. Slowly, the group rolls rounds of tree trunks into a circle, then eats a collective spaghetti dinner prepared by Dan and Cindy.

A new concern is voiced by Chad who observes that the Pacific fisher has now been listed and, since it dens in the spring, this could affect any version of fire during that season. There is discussion about the fallibility and limitations of the single study that has tried to determine smoke effects on fisher dens. Ever the fixer Joe notes that there are workarounds possible without shutting down all spring burns or rewriting the 2005 fire plan. But it is yet one more constraint on a program that struggles to let natural ignitions evolve and expire on their own, let alone factor in the historic burning from people.

It is an awkward discussion, orchestrated not around a fire full of crackle and warmth, but the fluorescent glow of an electric light. On August 31 the park had imposed stage-two fire restrictions, which prohibit open flame below 8,000 feet. That includes the Illilouette Creek campsite. We sit around a battery-powered lantern with an LED bulb. Some of the best fire scientists, fire management officers, and wilderness planners in California are unable to light a fire within a stone campfire ring around which they can converse about how to reintroduce free-ranging fire to the Illilouette basin, and beyond.

CHAPTER 7

The Trek

Back to the Future

Pack strings are scheduled to arrive at the Illilouette camp between 9 a.m. and 10 a.m. The trekkers are neither anxious to stay nor eager to leave. They have heard and seen what they came to hear and see, and they have said what they wanted to say. In the early light coming through the fir and pine, they sip coffee on stumps and eat breakfast, amid a steady chatter like the hum of mosquitoes. One by one, they roll up sleeping bags, tents, and pads.

There are no talks planned. It's a day for spontaneous conversations among twos and threes as we hike out, each with our own thoughts at our own pace.

• • •

The Illilouette basin is an island within an island amid a flood tide of megafires. The fires have gotten larger and meaner, burning outside the contours of recorded history, of fire behavior formulas, and of ecological models. The first two decades after the program began brought an order of magnitude increase in fire size and severity; and over two double decades, two orders of magnitude. Fires that stunned in the 1970s at 3,000 acres swelled to 30,000 and then to 300,000. In 1970 the largest category of fire on reporting forms was Class G, over 5,000 acres.

But size alone is a deceptive metric: these years also brought the greatest damages in terms of houses lost, public health compromised, and ecological losses. They were fires that fused the huge with the severe. Of California's 20 largest wildfires, 9 have burned since 2020, 14 have burned since the drought started in 2012, and 17 since 2000. Of its 20 most destructive, 15 have burned since 2012, and 17 since 2020, incinerating 43,411 and 47,881 structures, respectively. The damages have put the state's primary power utility, PG&E, into bankruptcy and left its executives facing criminal charges. The fires have taken out roughly 1 in 7 of the world's mature sequoias. They have decimated the population of Joshua Trees. Their smoke palls have driven metropolitan Californians indoors. Two fires burned across the Sierra Nevada, the Dixie fire twice. They testify to a century of a failed doctrine of total fire exclusion. Legislative bills before Governor Gavin Newsome—who survived a recall election while the trek was on—seek to reform liability law to boost prescribed fire, ramp up fuel treatments and fire-harden communities, and improve capability. "All lands, all hands, all options" runs the official mantra.[1]

Around Yosemite big fires have banged at the gates like barbarian hordes. The A-Rock and Steamboat fires (1990). The escaped Big Meadow fire (2009). The Rim fire (2013), which burst into the park for 77,000 acres. The El Portal and Dog Rock fires (2014). The Railroad and South Fork fires (2017). The Ferguson fire (2018). The Creek fire (2020) that started on September 4 and was not contained until Christmas Eve, in which a hundred trapped civilians escaped through a dramatic helicopter rescue, and for a while Wawona and its sequoia groves appeared at risk, and whose smoke so thickened in the valley that it exceeded air quality standards which passed from hazardous into downright dangerous and forced another closure on the park. The Merced and Tuolumne sequoia groves have escaped multiple threats by a whisker. Nationally, a legacy of fire exclusion, drought, logging, exurban sprawl, and dead trees has replaced the Big Blowup with a Big Blowback.

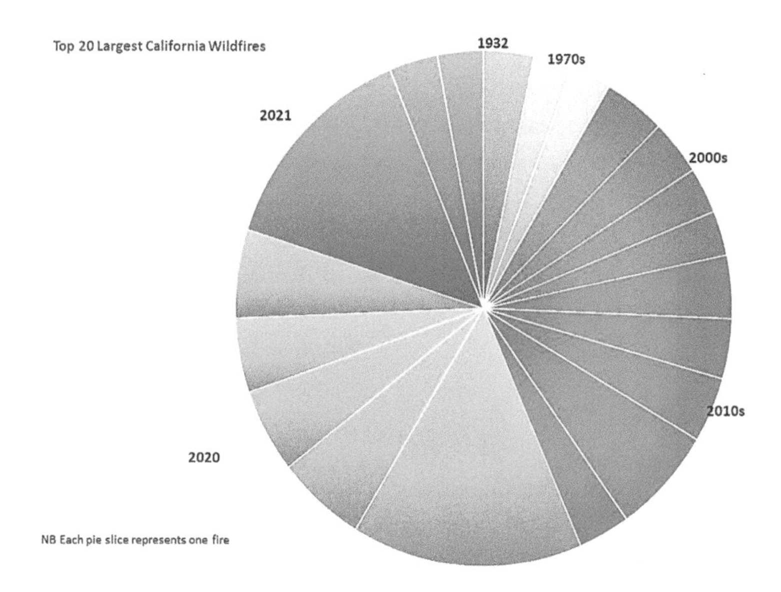

FIGURE 28 California historic fires by year and size.

In a guest essay for the *New York Times* Susanna Meadows compared the Yosemite of 2021 to what she remembered as a park ranger intern in 1993. The western entry to the park along California 41 seemed a postapocalyptic landscape of dead trees, trees burned to matchsticks, and "pathetic wisps" of waterfalls. The Sierra's last two glaciers were in rapid retreat. Backcountry streams, even in early July, were dry. Both Yosemite, and what she remembered as an "Edenic summer" amid an almighty nature, were, compared to her memory, devastated. The power dynamic between nature and humanity had shifted. Her son was impressed by the spectacle, but she recalled what was missing, and lamented that he would not see it as she had. Yesterday's green fire had met today's red fire and ended in a black tomorrow.[2]

Each generation remembers nature as it was imprinted in its youth, and each laments what has been added or removed from that founding template. Bertrand Russell once observed that the expression "return to nature," when traced to its origins, refers to the world an author knew as a child or in which they came of age. The first Americans entered the valley as though they had discovered a lost world; every account since then, however celebratory, is tinged with a shadow of declension, a different kind of lost world. Each entry is a revelation; each reentry, a reflection on loss.[3]

That may be true for humanity overall. We are a fire creature but an ice age creation. We emerged from a world shaped by ice into one increasingly shaped by fire. The entry to Yosemite has become a portal to the Pyrocene.

• • •

His 46 years in fire have taken their toll on Dan Buckley, Yosemite's fire management officer, though he manages to wear them well. From his start as a pickup firefighter at Sequoia-Kings to a CalFire engine to superintendent of the Arrowhead hotshots to tours at Yosemite (2000–4) and Sequoia-Kings (17 years in total) to fire director for the National Park Service at the National Interagency Fire Center, he has seen it all. Except he hasn't. "I'm seeing stuff now," he says, "I've never seen before." The fires have changed. The whole fire world has changed.[4]

Yosemite has always been special and has, among the national parks, always hosted an elite program in fire. But what can start can also stall. Missions change, execution both leaps and stumbles, privileged programs trust to luck. Between the Big Meadow blowup, California's punishing drought and outbreak of serial conflagrations, the fire organization's "Purge" and a general degrading of public assets nationally, Yosemite no longer had the fire program it needed. There were too many challenges and perhaps too little felt urgency to reform. Suppressing wildfires and not starting new ones slowly reclaimed default settings. Dan Buckley was recruited back to the field for a three-year tour of duty to rebuild the program.

The critical factor, he maintains, is always people. The program had lost its fire ecologist. It had no fuels specialist or active prescribed fire bosses. Resource Management had shed its vegetation manager, the person who would oversee mechanical treatments on encroaching woods and threatening invasives like cheatgrass. Yosemite seemed barely able to meet traditional issues with visitation and concessioners. It had little bandwidth to deal with a functioning but not really achieving fire program. It was pressed to keep bad fire out and good fire in. When Dan arrived, so did covid.

He hadn't seen it all, but he had seen far more than most of his peers, and he had a clear vision of what a fire program at Yosemite should be. The lapsed positions were filled; the newcomers would join the trek as part of their initiation. He would reboot the three theme-areas of Yosemite fire with the goal of moving from restoration to maintenance. The valley would have a full-immersion baptism by prescribed fire over the course of several years, then pivot to maintenance burning on a cycle of 4–12 years. The grove would expand treatments beyond its immediate circle of giants, which the staff regards as the hole in a dangerous donut. The surrounding forests need attention, a conviction confirmed by the 2020 Castle fire in the Sierra National Forest which blasted the long-standing premise that giant sequoias don't burn. They can, if the adjacent woods can put enough heat and fire into their crowns. The donut's hole has to expand to encompass the whole, and the Mariposa project desperately needs to be carried into the Tuolumne and Merced Groves, in what promises to be tough-love ecology amid unprecedented numbers of dead trees killed by drought, insects, disease, and past wildfires. What has traditionally been concentrated in the groves needs to expand to form a benign blackline along the park's exposed perimeter. But he reserves special attention for the wild.

For over a century the philosophy of wildland fire control has been to put out all small fires so they cannot become large. That dictum, he believes, needs to be reversed: we need small, good fires so that they can retard big, bad ones. New starts are innocent until proven guilty. In the wild every fire put out is a problem put off. If the park is to get anything like the 16,000 acres of good fire it would like yearly, it will need to do more with the opportunities nature provides. It will need to consider a relight program to restart ignitions that were extinguished for various reasons mostly connected with seasonal timing. It may need to more actively manage allowed starts, not simply let them passively fill out approved boxes drawn on maps, but gently push and pull them to speed up their residence time, lessen smoke, and make up for acres lost in the past. Whether such tactics can pass muster with wilderness character is unclear, though it leaves the wild neither untrammeled nor fettered.

Dan is halfway through the rebuilding. His tour is in many ways the culmination not just of a career but of a life spent around fire. He understands fire in the office as well as the field, he appreciates pyropolitics as much as pyrogeography. In some respects, his co-arrival with covid makes a useful analogy. If you don't vaccinate, wear masks, and social distance, covid will do the job for you. If you don't restore fire through drip torches and loose-herded natural ignitions, feral flame will do it instead. He can isolate valley and groves with special protective measures and help hold off the park from the rising tide of megafires, but he can't wall off fire-catalyzed climate change. The Yosemite celebrated as a legacy of Pleistocene ice is morphing into a Yosemite of Pyrocene fire.

The disadvantages are real. It takes constant effort to push a fire program in new directions: any slacking lets the maintenance slide and suppression creep back into dominance. The problem becomes someone else's, yet another unacknowledged debt passed on to future generations. The lightning fires extinguished in the Illilouette earlier that summer Dan describes as "like sacrificing my children." Every 30 years or so Yosemite has had to recharter its fire program, though the tempo seems to be picking up.

Still, there are unquestioned advantages open to someone who knows how to seize them. Yosemite is a flagship park and a celebrity landscape. It has better funding for fire than any other unit in the system. Compared to the Sierra and Stanislaus National Forests, Yosemite has more favorable natural conditions and a more forgiving public. It has less low-elevation grassland and shrubland

and more high-elevation granite. It has a spotty legacy of logging—part of land swaps to remove inholdings—but it missed the extensive clear-cutting that the national forests inherited, along with the tangle of regrowth that fire exclusion subsequently allowed to blossom. Dan has managed to fill that lapsed staffing. During the May evening burn in the valley, the superintendent herself enthusiastically worked a drip torch. Some 75–80 percent of all prescribed fire in the national park system occurs at Everglades and Big Cypress in south Florida. Glacier National Park has absorbed so many managed wildfires (every entry to the park passes through a major burn) that it may have reached a saturation point. If Yosemite can't restore fire, it's hard to imagine other, less illustrious western parks doing it.

Three days before the trek, Yosemite had three lightning fires. It suppressed the Ackerson Creek fire at 10 acres, the Tuolumne fire at 0.1 acre, and the North fire at 0.5 acres. All lay in the footprint of the Rim fire, all were stoked with heavy downed fuels. The Sierra National Forest attacked another start not far from the park border. Sequoia-Kings Canyon National Park had two—the Colony and the Paradise—between Ash Mountain and Lodgepole. They remained uncontrolled when the trek began, and by the time the trekkers had hiked back to Horizon Ridge, the two starts had merged into the KNP Complex, Ash Mountain was on alert for evacuation, Giant Forest had survived a flank of the fire that backed through at night, and CalFire was bulldozing lines outside the park boundary.

A week later Dan Buckley was dispatched to the KNP Complex as agency representative to join the Southwest Area Type I Incident Management Team 2. Fires that a century before had been successfully suppressed with saddle blankets, pine boughs, rakes, and wooden matches now exceeded the combined firepower of airtankers, helicopters, bulldozers, engines, chain saws, pumps, and the elaborate support system of infrared mapping planes, satellite uplinks for computers, thousands of Nomex-clad and specialty-tool-equipped crews, hardened by serial deployments and lengthening fire seasons. Now, crews prepped sequoia groves and buildings, doing under emergency conditions what decades of steady maintenance should have accomplished. By October 4, with the complex approaching 76,000 acres and a scant 11 percent contained, they were burning out along the length of the Generals Highway and hoping to gently back fire down Redwood Mountain and through Whitaker's Forest.

It seemed as though, almost 60 years after the fire revolution had begun, its western hearth was ready to report on the consequences. Satellite imagery recorded the evolution of a massive pyrocumulus punching through the troposphere. Winds at 55 miles per hour powered slopovers across the highway. A falling tree crushed a vehicle; another sent four firefighters to the hospital. Until the fire was well out, and snag fields were deemed safe to enter, officials were reluctant to report on the number of giant sequoias killed, but reckoned they were many. Farther south, estimates from the Windy fire suggested "hundreds" of mature sequoias—a tree long believed immune to fire damage—had died in high-intensity blowouts. Early reports suggested that between 2020 and 2021 some 13–19 percent of the Earth's population of mature sequoias had perished by fire.[5]

As the losses became public, Steve Botti, Yosemite's chief of resource management in the 1980s, and Tom Nichols, its fire management officer from 2002 to 2005, later regional fire director, and then NPS national fire director, authored a retrospective on the fire revolution. What Harold Biswell, Starker Leopold, Jan van Wagtendonk, Bruce Kilgore, and the other founders had warned might happen was happening. They itemized the endless impediments that always seemed to compromise the effort to restore fire. Fighting fires had bottomless support. Lighting them to advance ecological goals—fuel mitigation among them—depended on a special alignment of the stars and planets, and mostly personalities. (Tony Caprio in Sequoia-Kings once remarked that he went to work every day knowing that a change in wind might get him arrested.) Even prescribed fire qualifications required significant experience in suppression. Botti and Nichols regarded the fire revolution as, if not a failure, then a grand disappointment.[6]

Nichols observed that Sequoia-Kings held 11,000 acres of sequoias. Over 50 years that meant treatment of 220 acres a year. Instead of steady restoration the park was scrambling to apply ad hoc protection—draping aluminum foil around trunks and raking away woody debris—while wildfires encroached. The loss of thousands of sequoias he regarded as "the greatest resource management failure in the history of the NPS." The sequoia is on the NPS logo, and its cones on official hatbands. There is, he lamented, "no more iconic NPS resource than the sequoia."[7]

After his tour on the KNP Complex, Dan Buckley resolved that "it is likely that parts of the fire were more beneficial than detrimental. Yes, there were irreplaceable monarch sequoias and old growth lost and that is tragic but many survived and now the park has an excellent anchor point to conduct burns almost all throughout the park. Hopeful." The fire was doing what the Ferguson fire along the Merced River and the Ackerson complex that spilled over from the Stanislaus National Forest had done for Yosemite: providing a temporary protective cordon along the boundary and improving opportunities for deliberate burning. Yet it was an odd kind of hope that pivoted on what might have been worse and that relied on fortuitous conflagrations to do what nearly six decades of reforms at flagship fire parks had not been able to achieve.[8]

Six weeks later the UN Climate Change Conference (COP26) met in Glasgow to accelerate action to contain greenhouse gas emissions. Managing fires in living landscapes and fires burning lithic ones were at the heart of that challenge.

• • •

At the Mono Meadow Trailhead the trekkers sort through the gear hauled out by the pack strings. They exchange final words. For some it has been an initiation, for others an informative retreat, for all a renewal. For three days they could talk about and contemplate fire, not scramble to light it, fight it, or shuffle papers about its management. The fires indulged in no such breaks. Whether flaming or missing, whether burning living landscapes or combusting lithic ones, they were shaping Yosemite, the good fires wrestling with the bad.

• • •

The fires will be there with or without any fire officers, wilderness planners, ecologists, resource managers, or chiefs of staff. They just are, and as the sum of humanity's fire habits metastasizes, they are becoming bigger, meaner, and more feral. They are replacing the Earth's forecast future ice age with a runaway fire age from which there is no escape from flame, smoke, greenhouse gases, or combustion's collateral landscapes, or the source of the disruption, ourselves.

Leaving the park, the scene is less of glacial sculpting than of smoke in the valley and beyond the park, of burned woods and towns. Pass by Foresta and El Portal where serial fires have fashioned a fire corridor and peeled back the vegetation. Pass by streams of autos and semis burning gas and diesel. Pass by PG&E crews rebuilding power lines. Pass by industrial combustion shaping how people get to, interact with, and understand the park. Pass into the brown haze of smoke in the central valley, where people live in a third nature informed by the passage of lithic landscapes as they emerge from the geologic past, are combusted, and then disperse their effluent into the geologic future. That sky was also absorbing the emissions from coal-fired power plants in China and India, from tropical peatlands in Indonesia cleared and burned to make palm oil plantations, from Amazonian rainforest felled and burned to create pasture. Humanity no longer passed between two fires but between three, a genuine three-body problem for which there is no exact solution, only a negotiated transit.

Yosemite goes through annual cycles of water and fire, of spring melt and summer fire. Now those cycles have assumed the scale of millennia, perhaps of epochs. Over the past 2.6 million years Yosemite has passed through a succession of ice ages. Over the past 10,000 it has passed into a deepening fire age. Mountains that once bowed under the weight of ice now rise with flames. Valleys that once filled with glaciers now flood with smoke.

The Earth's latest interglacial is quickening into a mature Pyrocene. Fire has shed its modest immersion in terrestrial life and become an informing principle for the planet. Even those who dream of leaving a cindered Earth for a new world will do so on plumes of flame.

Epilogue
Between Three Fires

This was my sixth encounter with Yosemite.

I first saw it on a family vacation in 1957, when I was eight. We had moved to Phoenix, but both my parents were from the Bay Area and their extended families were still there, save for an outlier in Long Beach. We toured Sequoia-Kings Canyon, staying in the Grant Grove campground where my mother had been born 35 years earlier (she spent her first night in a suitcase), then to Bass Lake (a favorite of my father), and on to Yosemite where we saw bears, waterfalls, and the Glacier Point Firefall, before continuing to San Francisco, Redwood City, and Palo Alto for rounds with the relatives. Carbon dioxide load in the atmosphere was 315 ppm.

I returned twice while I was an undergraduate at Stanford, once with my roommate, Glenn, and my brother, Jim, and in a last hurrah, a field trip for a course on the geology of California. Glenn, Jim, and I hiked into the Little Yosemite Valley. A geologist with the USGS lectured on the succession of pluton emplacements that had massed into the Yosemite batholith, dates and data that washed over me like Yosemite Creek over the valley rim. That was in 1970, when California endured a historic outbreak of fires, and 1971, when the Illilouette Creek project was primed to accept its first fire and

I was headed to my fifth season with the North Rim Longshots. Atmospheric CO_2 was 325 ppm.

In 2000 the MacArthur Foundation Fellowship program gathered a group of six fellows for a short seminar in Yosemite Valley to discuss new thoughts on the origins of ice ages. I assume I was invited because of *The Ice*, my meditation on Antarctica. Stephen Schneider talked about climate. I enjoyed some nice chats with Mike Malin, mostly about Mars. We came, we talked, we left. I completed my first global survey, *Fire: A Brief History*. CO_2 was 365 ppm.

Ten years ago, while on a road trip as part of a fire reconnaissance of California, I spent a day at Sequoia-Kings and collected documents and chatted with people in Yosemite's various fire management offices. Kelly Martin, focused and impressive, was particularly helpful. I eventually wrote an essay about the Sierra parks' contribution to the fire revolution. CO_2 was 380 ppm.

The trek to Illilouette Creek completes the chronicle. I was 72 when it ended. California was experiencing a decade of drought and serial conflagrations; the six largest fires in California's recorded history had occurred in the past three years. The smoke in the San Joaquin Valley was so dense it was impossible to see the Sierra Nevada until you reached the foothills. CO_2 was 413 ppm.

At each encounter I saw Yosemite and California, and fire and ice, differently. My first visit was a family vacation that had all of us singing "California, Here I Come" as we drove pre-interstate highways. My last had me pondering the immense presence of California nature—its elemental power latent in rock, water, woods, and fire. It made real the notion of the sublime as a fusion of awe and terror.

And it tracked my relationship to fire. What began as branch-fed campfires had morphed into lightning-kindled snags on the Kaibab Plateau and finally into a geologic epoch. The sum of humanity's fire practices—quickened by how a fossil-fuel society shaped the way we lived—had morphed into the fire-informed equivalent of an ice age, a full-immersion baptism into a fire age. In that evolution it can stand for Earth overall.

• • •

Yosemite has always been special. Our first national park, a stele for monumental scenery, the loci for John Muir's and Ansel Adams's genius, a reference point for the Leopold Report, a pioneer of fire policy, and, less happily, a demonstration case for the difficulties of implementing that policy. It was still easier as a default setting to take fire out than to put it in, to sacrifice fires in living landscapes than impose limits on burning lithic ones. Progress depended less on policy than on personalities. Even natural fire relied on people, a paradox so elemental it can be embarrassing to state out loud. Yosemite showed what it takes to live with fire in contemporary wildlands and how hard that is to achieve.

Progress slowed, then stalled, while California spiraled into a witch's brew of drought, bug kill, and megafire that made any response other than urban-style firefighting suspect. Callouts to support endless firefights stripped the park of fire personnel. Yosemite felt the effects of four decades of a political ethos that promoted public squalor and private splendor; unless it could offer an initial public offering, it seemed even a crown-jewel park would have to scrimp and scrape.

Projects stumbled, people left. The park skidded through a period of black ice. Now it was rebuilding. It had a new fire management officer, a new prescribed fire specialist, and a new fire ecologist. Serious mechanical treatments were proceeding at the Mariposa Grove and along Garnet Ridge, with more planned for the Merced and Tuolumne Groves. Even amid yet another round of California conflagrations, the park had by October 2021 navigated its way through prescribed fires in the valley, pile burns around the Tuolumne and Merced Groves, and 50 lightning fires in the wild, each evaluated on its merits for suppressing, monitoring, nudging, and herding.[1]

FIGURE 29 Prescribed fire in Yosemite Valley, 2021.

Since the ice left, fire at Yosemite has been a fugue of people and nature. The prevailing fire regimes, especially in valley and grove, have been significantly anthropogenic. They were anthropogenic when the Ahwahneechee burned in the valley; they were anthropogenic when the cavalry, the CCC, and the Park Service stopped the burning; they were anthropogenic when the National Park Service partially restored the burning; they were anthropogenic when, thanks to industrial combustion, global warming encouraged wildland fires to go feral. They were anthropogenic when Miwok and shepherd burned the backcountry, and when, under a doctrine of the wild, scientists and fire officers allowed some lightning fires to burn more or less freely. They will be anthropogenic as global change broadens the range of effects and quickens the tempo of en-

FIGURE 30 Piles ready to burn along Garnet Ridge, part of a protective cordon around the Merced Grove.

vironmental disruption, and forces people to reconsider, once again, what mixture of fire lighting and fire fighting meets their sense of the place.

But those jostling regimes speak not just to people in general but to particular people in particular places and times. The right person could make, break, stall, nudge, tweak, or disrupt fire programs in ways that could last for decades, maybe centuries. The artifice that shapes Yosemite's fire program has as its nuclear core what William Faulkner famously proclaimed was the essence of art, the human heart in conflict with itself.

• • •

Joe Meyer, the park's chief of staff, and one of the trekkers, believes Yosemite is entering "its golden age of fire management." At all levels of park administration, fire is regarded as a critical issue, if not yet an existential one. The planets and stars—or in less metaphorical terms, the terrain, forest, knowledge, politics, and people—are aligning.

Today, Yosemite has marvels of fire management to match its granitic monuments. The valley has some serious black, a substantial down payment on its fire debt. The Mariposa Grove pioneered how to reintroduce fire to an iconic site. The park's lesser-known groves are undergoing felling, piling, and masticating prior to burning. The Illilouette basin belongs with a handful of American landscapes more or less approximating natural conditions. Yosemite's early efforts along its western perimeter, plus wildfires, helped shield border communities and checked some of the rampage of the Rim and Ferguson fires after they breached that boundary. (The 1996 Ackerson complex, which burned 47,000 acres in the park, was a dress rehearsal for the 2013 Rim fire that reburned most of its footprint.) These are, as the park proudly proclaims, "beacons" for other places.

Yet Yosemite is not where it wants to be. Its success is relative—outstanding in comparison to other places but not to what a flagship fire park might be. The valley has years, perhaps decades, if present work continues unabated before it enters a maintenance state, whatever maintenance might mean. The Mariposa Grove is robust, though the park was still cleaning up after the massive blowdown of January 19, 2021 (which even toppled 15 mature sequoias); but it is a sanctuary amid a latent storm of untreated fuels, some dating back to the nineteenth century, in which a worsening climate and megafires massing outside the gates could overwhelm its perimeter of sentinel sequoias; the Tuolumne Grove is far behind in treatment, and the Merced Grove is openly regarded as a biotic time bomb in desperate need of defusing. The Illilouette basin has granite for half of its perimeter, a model landscape not easily transported to the park's west side, much less exported to the Oregon Coast Range or the Southern Rockies.

As the saying goes, it's better to be lucky than good. Yosemite has been lucky enough to dodge the worst blights and conflagrations, and it's been good enough to have fashioned a firescape more resilient than that of its neighbors. Its 50 years of effort make it a paragon of the fire revolution, even if deteriorating conditions—none of them directly amenable to the fire program—mean those achievements may not be enough for the next 50 years. The question is no longer how to live with fire, but how to live with a full-blown fire age.

To those who kindled the fire revolution at Yosemite their task never seemed simple. They struggled against funding that favored suppression, fire cultures for which fighting was instinctive and lighting suspect, uncooperative weather and legacy fuel loads, an inadequate, inchoate science, a reward system that tolerated success but punished failure. People lost fires, people lost careers, they lost lives, and during the 1980s they lost a decade. Yet conditions today are more savage and implacable; the Leopold Report's vision to restore presettlement conditions looks naïve, if not illusory, and its reinstated fires an ignis fatuus. What worked in the 1960s and 1970s pales when confronted with the seemingly indominable forces that

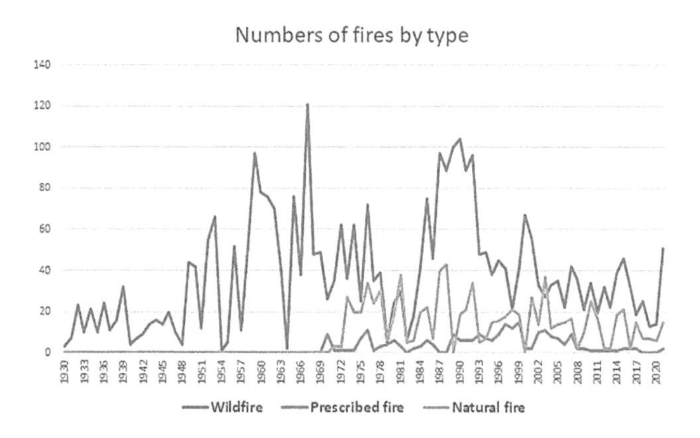

FIGURE 31 Yosemite's fire profile: numbers of fires by type (1930–2021).

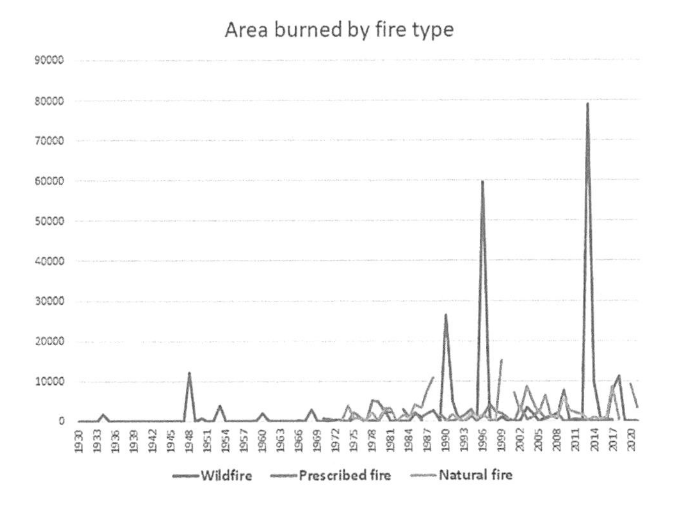

FIGURE 32 Yosemite's fire profile: burned area in acres by types of fire (1930–2021).

are blasting forests the length of the Sierra Nevada, like ecological earthquakes rupturing along the biotic equivalent of a San Andreas Fault. The park cannot rely on emergency mitigation during blow-ups and the burn scars of past megafires to substitute for systematic programs. To survive the next 50 years the park will need to be more good than lucky.

Yosemite matters. It's too big, too well known, too historic, too significant to the National Park Service and the American fire community, to be allowed to fail. Its burning beacons are tiny lights in a mountain range, in a country, and with a wider aperture in a planet, of growing ecological darkness. But they matter all the more for that, for whether the future comes as a golden age or a leaden one, it will surely be an age of fire. Amid a deepening Pyrocene we'll want those beacons of good fire to help guide us through the shoals of what promises to be a rising sea of bad ones.

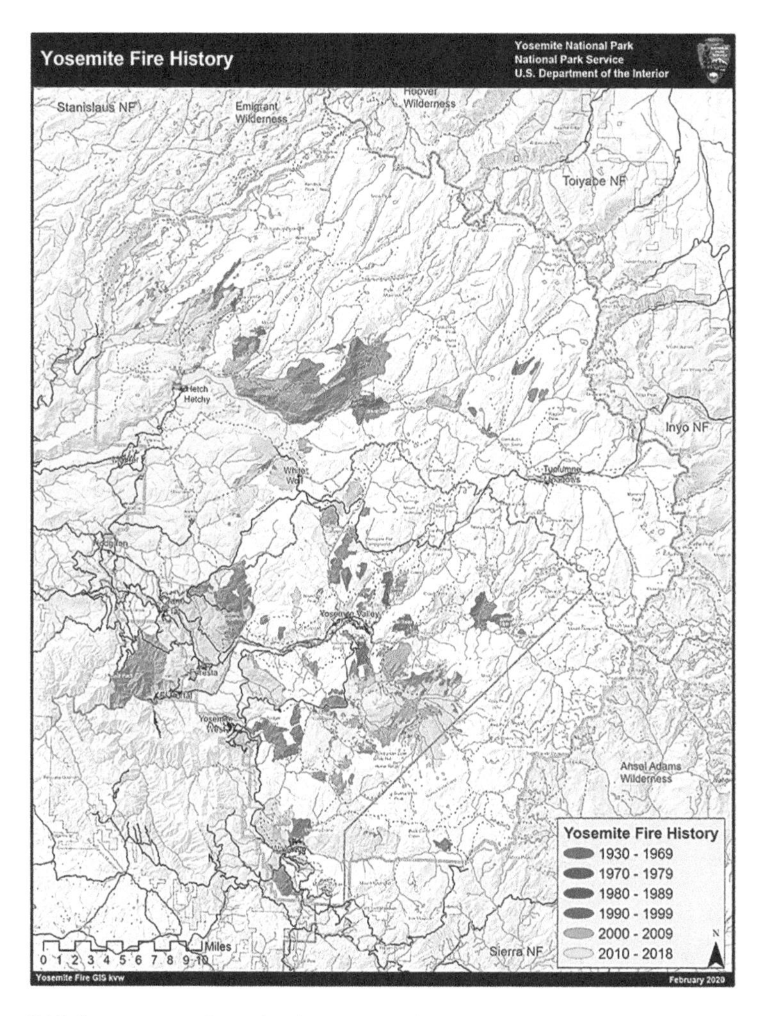

MAP 3 Yosemite fire atlas (1930–2021). Shaded areas to west of the park include the Rim and Ferguson fires.

Codicil

On the afternoon of July 7, 2022, a fire of human origin, cause unknown, started in the southern end of the Mariposa Grove. By evening it was 200 acres; by the morning of the 10th, 1,591 acres, of which 350 were in the grove proper. The grove and Wawona were evacuated.

The weather toughened into hot and dry, but—a bit of Yosemite luck—winds remained light. The fire moved north, mostly paralleling the western perimeter of the grove before arcing eastward along its northern boundary. Spotting was consistently reported several hundred yards ahead of the flaming front. Crews wrapped the Galen Clark cabin with foil, raked around the bases of giant sequoias, and erected sprinkler systems. An air tanker reported branches falling from the convective column. Containment was 0 percent.

By the morning of July 12, the Washburn fire had spread across 3,221 acres, held on the west by burnouts along Highway 41 and a dozer line flanking Wawona, on the south by the Mariposa Grove, and on the north by the South Fork of the Merced River, beyond which the 2017 South Fork fire offered a burn scar likely to reduce intensity should firebrands loft across the gorge. The South Fork had previously stopped wildfires from advancing south; now it would help keep the Washburn from spreading north.

There hadn't been as many prescribed fires as the fire staff wanted, or as many recent ones as needed. Prescribed burns resemble flu

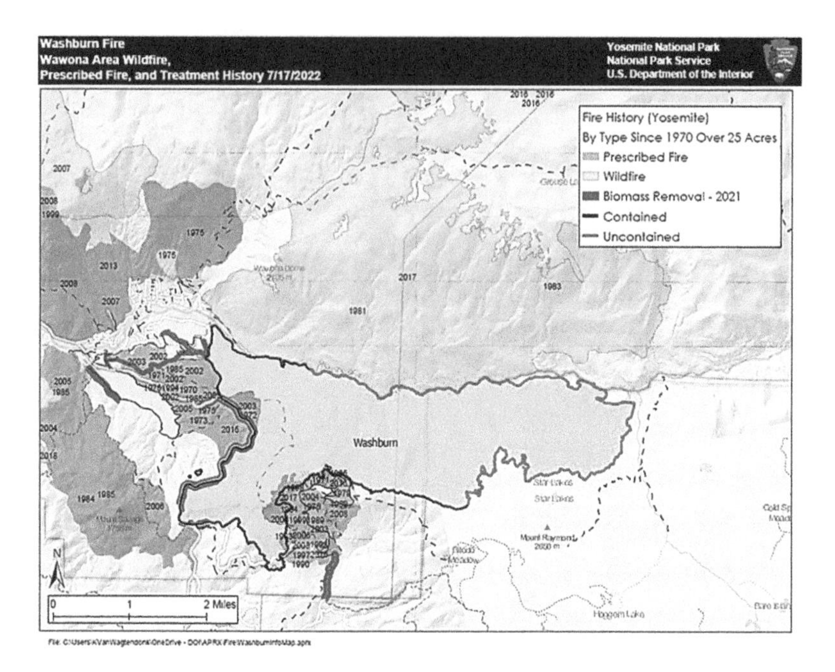

MAP 4 The Washburn fire, July 17, 2022. The legacy of previous treatments by prescribed fire, mechanical thinning, and managed wildfire shows clearly.

shots rather than vaccines for yellow fever or smallpox; they need to be redone frequently and are never as effective as hoped. But there had been enough treatments to allow crews to herd the flames around Wawona and the grove. Instead, the Washburn followed the available fuels: the advancing front burned into the dense donut (around the Mariposa Grove hole), a rugged landscape stuffed with trees from decades of fire exclusion, large numbers of which had been recently killed by drought, disease, beetles, and the January 2021 Mono-wind blowdown.

The front pushed east, left the park, and entered the Sierra National Forest. By July 16 the fire had grown to 4,822 acres, had 1,623 personnel assigned to it, and was 37 percent contained. Where the South Fork met Iron Creek, handcrews, assisted by over 300 drops from helicopters, promised to hold the eastern front. Burnouts, river, retardant, and granite pinched off further advance. By July 19 the Washburn was 4,863 acres and 58 percent contained. The next day containment reached 100 percent at 4,886 burned acres.[1]

A year later the lightning-kindled Red fire burned 8,410 acres within the Illilouette Creek watershed—only 537 acres smaller than the largest burn in the 50-year history of the restoration program. Some selective herding by hand crews held its southern flank within the basin. Its northern half included the lunch site visited on the second day of the trek.[2]

• • •

Sometimes coincidence segues into apologue. On July 16, 2021, after the Washburn fire had skirted the Mariposa Grove and Wawona, Jan van Wagtendonk passed away. Thanks to decades of treatment, along with emergency rehab after the blowdown, the Mariposa Grove suffered no worse than a handful of scorched giants—a fitting legacy to Jan, Bob Barbee, Doc Biswell, and the others of that intrepid cohort. The Washburn fire had spread exactly where nothing had been done over the past 50 years to retard it.

Meanwhile, the Earth Island Institute filed suit in federal court to end what it regarded as commercial logging along the Merced Grove Road, part of the restoration program for that grove. Cutting and hauling away the trees, the plaintiffs argued, would increase carbon emissions, worsen the climate crisis, and by removing trunks less than 20 inches in diameter reduce forest resilience. On July 1 a federal judge ordered the park to cease operations and to use only prescribed fire as outlined in its 2017 fire management plan.[3]

Touring the Mariposa Grove, Garrett Dickman, the park's vegetation manager, overseer of the Merced Grove project, and one of the Illilouette trekkers, thought the Mariposa Big Trees would survive. For now. "I'm an eternal optimist," he explained, "but there are things that are happening, they're just moving faster than any of us imagined. The thing I tell everybody is, go experience giant sequoias now.[4]

AUTHOR'S NOTE

Pyrocene Park is an extended essay, a nonfiction equivalent to a novella. I found Yosemite's fire story too rich for a short essay, yet a full-blown book was a bigger commitment than I was prepared to make, especially when covid prevented access to archives, and a longer manuscript would deny me the chance to use the Illilouette trek as an organizing device. I wanted a text as much personal reflection as scholarly monograph.

Yet the genre has its liabilities: it requires juggling the various pieces into a proper proportion. Emphasize one topic too much and it unbalances the whole, and leaves readers to expect a comparable scale of treatment for other issues. For example, the project begs for short biographies of the major figures, which would help anchor Yosemite's fire program with the people who have shaped it. (As Dan Buckley observes, it comes back to people.) Instead, my extended essay shrinks those bios into sketches. And again, the evolution of park policy had to reconcile local conditions with national policies and personalities; the National Archives has contributions as significant as those in Yosemite's archives. There are always trade-offs. These were mine, and the outcome seems to yield something like a narrative case study.

And the codicil? The Washburn fire, the Merced Grove lawsuit, and Jan's passing occurred while I was working through the copyedited manuscript. I couldn't ignore those events—they were too relevant, uncannily so, not to include. But the epilogue couldn't take another ending and maintain coherence; I had designed it to be what it was. A codicil seemed a good compromise. That's part of the awkwardness about writing history into real time. The ending keeps moving and threatens to unmoor the organizational logic of all that goes before. William Faulkner might be right about the past never being over, but the future is never just to come; it's always already upon us.

The motive force behind this project was Jan van Wagtendonk. Jan read my account of a pack trip to Banff that Parks Canada had organized when I was writing a fire history of Canada, and wanted me to adapt that style to a trek into the Illilouette basin. Unfortunately, I had health issues the first year, and then covid intervened the second, and when the trek was finally underway, Jan was terminally ill. But he was the organizing spirit behind the trek, my account, and much of the fire program they both celebrate.

Dan Buckley took a day out of his ever-busy schedule to host me and two companions, Jordan Fisher Smith and Maura O'Connor, on an insider's look at Yosemite's fire program. Jordan and Maura then joined me for a second day's excursion to Glacier Point and the Mariposa Grove. Our conversations were a spur to think more pointedly about what we were seeing and how to express it. Jennifer Anderson briefed me on the fuels and prescribed fire program. Athena Demetry arranged for me to converse with those resource management specialists who had some connection to the fire program; most accompanied the trek, but the early primer was hugely helpful to someone uninitiated into the nuances of Yosemite's complex landscapes, both natural and bureaucratic; my thanks to Garrett Dickman, Chad Anderson, and Nicole Athearn. Scott Stephens and Mark Fincher forwarded documents that proved vital to both my understanding and the text. Kent van Wagtendonk contributed some statistics and an invaluable suite of maps. A special acknowledgment goes to Joe Meyer for a fun dinner and chat about the

Park Service, even if, on the trek, he grimaced every time I pulled out my notebook.

Not on the trek but sources for excellent background information were Tom Nichols, Kelly Martin, and Robert Reese; they have my respect as well as my thanks. I had met Tom and Kelly earlier in their storied careers and was pleased to reconnect. It would have been better for all of us if I knew how to interview, but I learned my craft from documents and fly-on-the-wall listening, and my "strategy" (such as it is) is to get people to talk about fire in a way they are comfortable with. Especially on an operational level, fire has a rich, metaphoric lexicon and sharp rhythms that I never tire of hearing.

Critically, Paul Rogers, park archivist, sent me digital copies of many items in the Yosemite Park Archives. I had expected to do the archival searching myself, but shortly before I departed for Yosemite I learned that the archives were closed due to the revival of covid. Paul graciously offered to scan (within limits) those items I thought relevant based on finding aids. At one point Virginia Sanchez also contributed. Their assistance meant I could transform the project from the journal of a trip through the Illilouette into something like a trek through the backcountry of Yosemite's fire history.

Then, too, I can't overlook the three anonymous reviewers whose comments were polite, pointed, and punctuated with references I had not previously encountered but proved inestimable in revising the original text. The care they invested inspired me to want a manuscript worthy of their effort.

The University of Arizona Press deserves, once again, acknowledgment for taking on a manuscript that does not reside tidily within their catalog. Allyson Carter, Amanda Krause, Leigh McDonald, and Abby Mogollon were all critical—as always a pleasure to work with. And Kerry Smith applied his usual copyediting skills with more forbearance and patience than I deserve.

My thanks to all.

NOTES

Chapter 1

1. For the general fire practices of California's Indigenes, see Kat Anderson, *Tending the Wild: Native American Knowledge and the Management of California's Natural Resources* (Berkeley: University of California Press, 2013).
2. Though dated, a useful survey of Indigenous fire practices from a Yosemite perspective is available in C. Kristina Roper Wickstrom, *Issues Concerning Native American Use of Fire: A Literature Review*, Publications in Anthropology 6 (El Portal, Calif.: Yosemite Research Center, 1987). For a global perspective, see Stephen J. Pyne, *Fire: A Brief History*, 2nd ed. (Seattle: University of Washington Press, 2019), with special reference to pp. 46–64. The fire-history figure comes from Andrew E. Scholl and Alan H. Taylor, "Fire Regimes, Forest Change, and Self-Organization in an Old-Growth Mixed-Conifer Forest, Yosemite National Park, USA," *Ecological Applications* 20, no. 2 (March 2010): 362–80. On sightings of abandoned campfires, see John Muir, *My First Summer in the Sierra* (Boston: Houghton Mifflin, 1917), 204, 226.
3. Clark quote from Emil F. Ernst, "Forest Encroachment on the Meadows of Yosemite Valley," *Sierra Club Bulletin* 46, no. 8 (October 1961): 21–32. Other Clark observations from Galen Clark, *Indians of the Yosemite Valley and Vicinity: Their History, Customs and Traditions* (Yosemite Valley, Calif.: Galen Clark, 1904), especially pp. 31 and 49.
4. Alfred Runte, *Yosemite: The Embattled Wilderness* (Lincoln: University of Nebraska Press, 1990), 37. For data on how the new fire regime mostly continued the old one, see Scholl and Taylor, "Fire Regimes,"

which suggests for the lower Tuolumne Valley a shift from a median fire-return interval of 10 years to one of 13.

5. J. W. Whitney, *Report of the Commissioners to Manage the Yosemite Valley and the Mariposa Big Tree Grove, for the Years 1866–7* (San Francisco, Calif.: Towne and Bacon, 1868), 6. Frederick Law Olmsted, "The Yosemite Valley and the Mariposa Big Trees: A Preliminary Report," in *Landscape Architecture* 43, no. 1 (October 1952): 12–25; quote on p. 24; or available online at https://www.nps.gov/parkhistory/online_books/anps/anps_1b.htm.
6. Mills quoted in Runte, *Yosemite*, 52, 59–60.
7. Ernst, "Forest Encroachment," 29–30.
8. On the firefall, I follow Runte, *Yosemite*, 93.
9. Hal K. Rothman, *Blazing Heritage: A History of Wildland Fire in the National Parks* (New York: Oxford University Press, 2007), 18.
10. Regulations quoted from Rothman, *Blazing Heritage*, 19. On the role of Sargent and the Northeast generally, see Stephen Pyne, *The Northeast: A Fire Survey*, To the Last Smoke 7 (Tucson: University of Arizona Press, 2019), especially pp. 4–7, 117–24.

Chapter 2

1. A. E. Wood, *Report of the Acting Superintendent of the Yosemite National Park* (Washington, D.C.: Government Printing Office, 1891), 3.
2. Wood, *Report*, 1891, 8.
3. Wood, *Report*, 1891, 3–5.
4. A. E. Wood, *Report of the Acting Superintendent of the Yosemite National Park* (Washington, D.C.: Government Printing Office, 1892), 5–6.
5. Rothman, *Blazing Heritage*, 20–21.
6. Wood, *Report of the Acting Superintendent of the Yosemite National Park* (Washington, D.C.: Government Printing Office, 1893), 649; G. H. G. Gale, *Report of the Acting Superintendent of the Yosemite National Park* (Washington, D.C.: Government Printing Office, 1894), 675.
7. Gale, *Report*, 1894, 673.
8. Gale, *Report*, 1894, 675–76.
9. Lt. Colonel S. B. M. Young, *Report of the Acting Superintendent of the Yosemite National Park* (Washington, D.C.: Government Printing Office, 1896), 4; Lt. Colonel S. B. M. Young, *Report of the Acting Superintendent of the Yosemite National Park* (Washington, D.C.: Government Printing Office, 1897), 4–5.

10. J. W. Zevely, *Report of the Acting Superintendents of the Yosemite National Park* (Washington, D.C.: Government Printing Office, 1899), 1057, 1060.
11. Zevely, *Report*, 1899, 1056–57.
12. E. F. Willcox, *Report of the Acting Superintendent of the Yosemite National Park* (Washington, D.C.: Government Printing Office, 1899), 4–5; William Forse, *Report of the Acting Superintendent of the Yosemite National Park* (Washington, D.C.: Government Printing Office, 1899), 6–7.
13. At first blush the continuity in fire regimes amid the onslaught of gold seekers seems counterintuitive. But there are plenty of accounts that report a more or less unbroken chronicle of fire setting, and the study by Scholl and Taylor, "Fire Regimes, Forest Change, and Self-Organization in an Old-Growth Mixed-Conifer Forest, Yosemite National Park, USA," reckoned the median fire-return interval shifted from 10 years to 13, well within natural fluctuations and probably a reflection of sheep and fire interacting differently than deer and fire. Willis W. Wagener reviewed early studies (prior to 1961) and concluded that "no major change [in fire return intervals] occurred until after 1900, corresponding roughly with the establishment of the national forests"; see Wagener, "Past Fire Incidence in Sierra Nevada Forests," *Journal of Forestry* 59, no. 10 (October 1961): 747.
14. John Bigelow Jr., *Report of the Acting Superintendent of Yosemite National Park* (Washington, D.C.: Government Printing Office, 1904), 8.
15. H. C. Benson, *Report of the Acting Superintendent of Yosemite National Park* (Washington, D.C.: Government Printing Office, 1905), 10.
16. Charles Howard Shinn, "Work in a National Forest: No. 5—Holding Down a Mountain Fire," *Forestry and Irrigation* 13, no. 12 (December 1907): 639, 641, 643, 642.
17. Fernow quoted in Andrew Denny Rodgers III, *Bernhard Eduard Fernow: A Story of North American Forestry* (New York: Hafner, 1968; facsimile of 1951 edition), 167.

Chapter 3

1. Ben Minteer and I elaborate this theme in "Restoring the Narrative of American Environmentalism," *Restoration Ecology* 21, no. 1 (January 2013): 6–11.
2. John Muir, *Nature Writings* (New York: Library of America, 1997), 618.
3. Muir, *Nature Writings*, 41, 110–11.
4. Muir, *Nature Writings*, 163, 184, 408, 438–39, 404.

5. Quoted in H. H. Biswell, "Forest Fire in Perspective," in *Proceedings, Tall Timbers Fire Ecology Conference*, No. 7 (Tallahassee: Tall Timbers Research Station, 1967), 46.
6. Muir, *Nature Writings*, 715–16, 632, 534.
7. Muir, *Nature Writings*, 813, 632–33.
8. John Muir, "The New Sequoia Forests of California," *Harper's New Monthly Magazine*, November 1878, 822; Muir, *Nature Writings*, 431.
9. Gifford Pinchot, *Breaking New Ground* (Seattle: University of Washington Press, 1972), 44.
10. Muir, *My First Summer*, 226.

Chapter 4

1. Frederick M. Jones, *Report of the Acting Superintendent of Yosemite National Park* (Washington, D.C.: Government Printing Office, 1904), 10.
2. Gifford Pinchot, *A Primer of Forestry, Part I: The Forest* (Washington, D.C.: Government Printing Office, 1899), 77; U.S. Department of Agriculture, Forest Service, *The Use of the National Forest Reserves: Regulations and Instructions* (Washington, D.C.: Government Printing Office, 1905), 12.
3. A somewhat more nuanced review of Powell's views is available in Stephen Pyne, *The Interior West: A Fire Survey*, To the Last Smoke 6 (Tucson: University of Arizona Press, 2018), 5–10. On his published perspective, see J. W. Powell, "The Non-Irrigable Lands of the Arid Region," *Century Magazine* 39, no. 6 (April 1890): 915–22, p. 919. For Fernow's view, see Rodgers, *Bernhard Eduard Fernow*, 154.
4. Dated in its judgment but still valuable as an entry point into the literature is C. Raymond Clar, *California Government and Forestry from Spanish Days Until the Creation of the Department of Natural Resources in 1927* (Sacramento, Calif.: Division of Forestry, 1959); there are many references throughout the text but see especially 209–11, 320–43, 488–94. See also Stephen J. Pyne, *Fire in America: A Cultural History of Wildland and Rural Fire* (Princeton: Princeton University Press, 1982), 112–18. For a modern perspective, see Daniel May, "Taking Fire: The Historical and Contemporary Politics of Indigenous Burning in Australia and the Western United States" (PhD diss., Australian National University, 2020).
5. On the episode with the psychologist, see John Shea, "Our Pappies Burned the Woods," *American Forests* 46 (April 1940): 159–62, 174.

6. Quoted in Louis Barrett, *A Record of Forest and Field Fires in California from the Days of the Early Explorers to the Creation of the Forest Reserves* (San Francisco, Calif.: U.S. Forest Service, 1935), 48.
7. S. B. Show and E. I. Kotok, *The Role of Fire in the California Pine Forests*, United States Department of Agriculture Bulletin 1294 (Washington, D.C.: Government Printing Office, 1924), 47.
8. DuBois deserves a biography. I give a brief sketch of his career in *Fire in America*, 265–68. Coert duBois, *Trail Blazers* (Stonington, Conn.: Stonington, 1957), x, 76. Basic biographical information is available in "U.S. District Forester Resigns—Colonel Coert Dubois to Enter Consular Service," *Sausalito News*, October 25, 1919. For more on the project, see duBois, *National Forest Fire-Protection Plans* (Washington, D.C.: Government Printing Office, 1911) and "Organization of Forest Fire Control Forces," *Society of American Foresters, Proceedings* 9 (October 1914), 512–21.
9. Quotes from *Sausalito News*, "U.S. District Forester Resigns," and duBois, *Trail Blazers*, 79.
10. On Leopold, see Aldo Leopold, "'Piute Forestry' vs. Forest Fire Prevention," *Southwestern Magazine*, March 1920, reprinted in David E. Brown and Neil B. Carmony, eds., *Aldo Leopold's Southwest* (Albuquerque: University of New Mexico Press, 1990), 139–42. On Greeley's unseemly sneer, see William B. Greeley, "'Piute Forestry' or the Fallacy of Light Burning," *Timberman* 21, no. 5 (March 1920): 38–39, reprinted in *Forest History Today*, Spring 1999, 33–37.
11. On White's alternative path, see Rothman, *Blazing Heritage*, 41–43. On Grey, see Zane Grey, "A Warning to California," *Outdoor America*, November 11, 1924.
12. "Superintendent's Monthly Report for August, 1924," September 4, 1924, Yosemite Park Archives, 24–25.
13. "Superintendent's Monthly Report for August, 1924," 26.
14. For a quick summary of events, see Rothman, *Blazing Heritage*, 43–45. For context within the fire history of Glacier, see also Stephen J. Pyne, *Northern Rockies: A Fire Survey* (Tucson: University of Arizona Press, 2016), 115–24.
15. Rothman's *Blazing Heritage* is by far the richest account of fire management by the NPS, but Pyne, *Fire in America*, still provides some helpful context, 295–305.
16. Coffman's extraordinary career is granted good coverage in Rothman, *Blazing Heritage*, 53–68.

17. The best account of the CCC's fire-related efforts is still Pyne, *Fire in America*, 360–70. For the National Park Service specifically, see Rothman, *Blazing Heritage*, 60–73.
18. Rothman, *Blazing Heritage*, 59, 68. Rothman says "fire suppression" but I suspect the original text read "fire *pre*suppression," since "fire suppression" and "firefighting" are interchangeable categories.
19. Matthias Gafni, "An 800-Mile Firebreak Once Traversed California. What Happened?," *San Francisco Chronicle*, November 13, 2020, https://www.sfchronicle.com/california-wildfires/article/An-800-mile-firebreak-once-protected-15713546.php. The best account of the evolution of the 10 a.m. policy is Pyne, *Fire in America*, 272–87.
20. Coffman quotes from Rothman, *Blazing Heritage*, 68. On Yosemite Valley, see Runte, *Yosemite*, 175.
21. Coffman, "Board of Review Report: Rancheria Mountain Fire 1948," Yosemite National Park Archives, Protection Division Records, Box 1931–1975, 1–5.
22. Coffman, "Board of Review Report, Rancheria Mountain Fire," 3–5.
23. Rothman, *Blazing Heritage*, 80–81.
24. Herbert Kaufman, *The Forest Ranger: A Study in Administrative Behavior* (Baltimore, Md.: Johns Hopkins University Press, 1960); Francis Fukuyama, *Political Order and Political Decay: From the Industrial Revolution to the Globalization of Democracy* (New York: Farrar, Straus, Giroux, 2015).
25. For snapshot summaries of the Department of the Interior's fire buildup, see Stephen J. Pyne, *Between Two Fires: A Fire History of Contemporary America* (Tucson: University of Arizona Press, 2015), 134–58.

Chapter 5

1. Abbott L. Ferriss, *National Recreation Survey*, ORRRC Study Report 19 (Washington, D.C.: Government Printing Office, 1962), 287–88. William J. Robbins et al., *A Report by the Advisory Committee to the National Park Service on Research* (Washington, D.C.: National Academy of Sciences, 1963). Leopold Report reprinted in National Park Service, *Compilation of the Administrative Policies for the National Parks and National Monuments of Scientific Significance (Natural Area Category)* (Washington, D.C.: Government Printing Office, 1970). On the role of science in the parks, see Richard West Sellars, *Preserving Nature in the National Parks: A History* (New Haven,

Conn.: Yale University Press, 2009). And for a smart discussion of Starker Leopold and some of the collateral consequences of his fabled report, see Jordan Fisher Smith, *Engineering Eden: A Violent Death, a Federal Trial, and the Fight over Controlling Nature* (New York: Crown, 2016).

2. "Practically illegal" quote from Bud Moore, from Linda S. Mutch and Robert W. Mutch, "Wilderness Burning: The White Cap Story," unpublished manuscript with transcriptions from recordings from the anniversary event, courtesy of Robert Mutch.
3. National Park Service, *Compilation of the Administrative Policies*, 1970, 102.
4. National Park Service, *Compilation of the Administrative Policies*, 1970, 106–7.
5. Green Book's full title was National Park Service, *Compilation of the Administrative Policies for the National Parks and National Monuments of Scientific Significance (Natural Area Category)* (Washington, D.C.: Government Printing Office, 1968); quotes from p. 17. A good summary of the transition is available from Bruce M. Kilgore, "Origin and History of Wildland Fire Use in the U.S. National Park System," *George Wright Forum* 24, no. 3 (2007): 92–122. For much of the history outside his personal experience he relies on Rothman, *Blazing Heritage*. A still wider panorama is available in Pyne, *Between Two Fires*.
6. Bob Barbee, interview by Brenna Lissoway, October 29, 2008, *I Remember Yosemite: Yosemite National Park Oral History Project*, Yosemite National Park Archives, El Portal, Calif.
7. The most condensed documentation is an oral history interview with A. Starker Leopold conducted by Becky Evans et al., "Sierra Club Nationwide II" (Sierra Club History Committee, 1984), 11–12. See also the essay Starker wrote upon the return from his second Mexico trip: "Adios, Gavilan," *Pacific Discovery*, January–February 1949, 4–13. A monograph on the two Leopolds and their Mexican travels is available in William Forbes, "Revisiting Aldo Leopold's 'Perfect' Land Health: Conservation and Development in Mexico's Río Gavilan" (PhD diss., University of North Texas, 2004).
8. Evans et al., "Sierra Club Nationwide II," 12.
9. Evans et al., 12. For a snapshot summary of Sauer's overlooked involvement with fire, see Stephen J. Pyne, *Slopovers: Fire Surveys of the Mid-American Oak Woodlands, Pacific Northwest, and Alaska*, To the Last Smoke 8 (Tucson: University of Arizona Press, 2019), 27–30.

10. On Starker Leopold's continuing involvement with the park service and the consequences of his eponymous report, see Jordan Fisher Smith's highly engaging *Engineering Eden: A Violent Death, a Federal Trial, and the Struggle to Restore Nature in Our National Parks* (New York: The Experiment, 2019; reprint of 2016 edition).
11. Biswell published an early account of his conversion experience in "Man and Fire in Ponderosa Pine in the Sierra Nevada of California," *Sierra Club Bulletin* 44, no. 7 (October 1959): 44–53, and elaborated the outcomes in his professional memoir, *Prescribed Burning in California Wildlands Vegetation Management* (Berkeley: University of California Press, 1989).
12. Emil F. Ernst, *Preliminary Report on the Study of the Meadows of Yosemite Valley* (Yosemite National Park, Calif.: National Park Service, 1943), 17–18. See also Emil F. Ernst, "Vanishing Meadows in Yosemite Valley," *Yosemite Nature Notes* 28, no. 5 (May 1949): 34–41, and "Forest Encroachment on the Meadows of Yosemite Valley," *Sierra Club Bulletin* 46, no. 8 (October 1961): 21–32. The observation that some burning persisted up to 1930 comes from Kilgore, "Origin and History of Wildland Fire Use," 93, citing an unpublished report by Emil Ernst.
13. Barbee, interview, 35.
14. The finale of the firefall and the later riot are covered in Runte, *Yosemite*, 202. The postriot transfers come from Jan van Wagtendonk interview with the author.
15. Runte, *Yosemite*, 166, 165.
16. H. Thomas Harvey, Howard S. Shellhammer, and Ronald E. Stecker, *Giant Sequoia Ecology: Fire and Reproduction*, Scientific Monograph Series 12 (Washington, D.C.: National Park Service, 1980), provides a summary chronology and reviews both Hartesveldt's pioneering research as well as that which followed; timeline on xvii–xviii.
17. Harvey, Shellhammer, and Stecker, *Giant Sequoia Ecology*, 146.
18. Thomas W. Swetnam, *Tree-Ring Reconstruction of Giant Sequoia Fire Regimes: Final Report* (Tucson, Ariz.: Laboratory of Tree-Ring Research, 1992).
19. Swetnam, *Tree-Ring Reconstruction*, 85–86; Thomas W. Swetnam et al., *Fire History Along Elevational Transects in the Sierra Nevada, California: Final Report to Sierra Nevada Global Change Research Program, United States Geological Survey, Biological Resources Division, and Sequoia, Kings Canyon, and Yosemite National Parks* (Tucson, Ariz.: Laboratory of Tree-Ring Research, 1998), 39.
20. Barbee interview, 33–34, 40.

21. Barbee, 34.
22. Norman L. Christensen et al., *Final Report: Review of Fire Management Program for Sequoia-Mixed Conifer Forests of Yosemite, Sequoia and Kings Canyon National Parks* (draft, February 5, 1987). The larger controversies and the Christensen committees they sparked are discussed in Pyne, *Between Two Fires*, 199–200, 202, 235, 239, 243.
23. Christensen et al., *Final Report*, includes a seven-page digest, but the longer text is valuable for how it addresses the particular controversies involved.
24. Jan van Wagtendonk, interview by Stephen J. Pyne, 2011.
25. "Horizon Fire, Fire Situation Analysis, Part II, 6/20/94 Amendment," in Fire Management Records, 1930–2020, YOSE 232940, Series III: Wildland Fire Reports, 1970–2001, Subseries C: Major Fire Incidents, 1968–1999, Subsubseries 7: Horizon Fire, 1994, Yosemite Park Archives, El Portal, Calif.

Chapter 6

1. Quotes from "Fire Control Plan, Yosemite National Park, 1964," in Fire Management Records, 1930–2010, YOSE 1017, Series VI: Administrative Files, 1931–1998, Subseries A: Central Files, 1931–1996, Yosemite National Park Archives, El Portal, Calif., 5, 1.
2. "Report of the Board of Review, 1968 Fire Season, Yosemite National Park (Part 2 of 2)," in Fire Management Records, 1930–2010, YOSE 1017, Series I: Fire Incident Reports (Forest and Structural), 1931–1969, Yosemite National Park Archives, El Portal, Calif., 1.
3. "Report of the Board of Review," 164.
4. Jan van Wagtendonk, interview by Brenna Lissoway, April 13, 2010, *I Remember Yosemite: Yosemite National Park Oral History Project*, Yosemite National Park Archives, El Portal, Calif., 2011.
5. See the biographical summary in Lesley Ragsdale and Desiree Ramirez, "Finding Aid," Jan van Wagtendonk Collection, 1923–2007 (bulk dates: 1972–2000) (September 7, 2012), Catalog Number YOSE 233112, Yosemite Collection Number 1015, and more fully in van Wagtendonk, interview, 2011. Other information from personal interview by Steve Pyne in May 2011.
6. Quote from van Wagtendonk, interview, 2011. On Biswell as artist, see van Wagtendonk, interview by Renna Lissoway, 2010, 34–35.
7. Jan W. Van Wagtendonk et al., *Fire in California's Ecosystems* (University of California Press, 2006). The book was expanded and published by the University of California Press in 2018.

8. Forest Service study cited in Ernst, "Vanishing Meadows in Yosemite Valley," 40.
9. An effort to recover the backcountry history exists in Jim Snyder, "Wilderness Historic Resources Survey 1990–1995," unpublished report in Yosemite Research Library. The insight that the park's goals restructured not only human access to Yosemite but its history is a theme of the study. I'm indebted to Mark Fincher for sending me a copy of this document.
10. "Wilderness Historic Resources Survey: Report on the 1993 Season," in Snyder, "Wilderness History Resources Survey 1990–1995," 4–6.
11. "Wilderness Historic Resources Survey: Report on the 1993 Season," 7.
12. Robert D. Barbee and Harold Biswell, *Environmental Restoration Program for Yosemite National Park: 1970*, Protection Division Records, Box Prescribed Fire 4–14, 1969–1985, Folder Environmental Restoration Program 1970, Yosemite Park Archives.
13. "Prescribed Fire Management Plan: Yosemite National Park, 1976," Jim Bennedict Collection, Box 5b 1968–1990, Folder Fire Management Plan 1979, Yosemite Park Archives, 2–3, 7.
14. Stephen J. Botti and Tom Nichols, "The Yosemite and Sequoia-Kings Canyon Prescribed Natural Fire Programs, 1968–1978," Jim Benedict Collection, Box 5b, Folder Prescribed Burning 1968–1978, Yosemite Park Archives, 6.
15. Botti and Nichols, "Yosemite and Sequoia-Kings Canyon," 7, 10–11.
16. Stephen J. Botti, *Natural, Conditional, and Prescribed Fire Management Plan: 1979*, Jim Benedict Collection, Box 5b 1968–1990, Folder Fire Management Plan 1979, Yosemite Park Archives, 35–36.
17. "Memorandum to Superintendent, Yosemite National Park from Chief, Fire Management, Subject: Draft Natural, Conditional, and Prescribed Fire Management Plan, Yosemite, 1979," April 27, 1979, and "Addendum 1981: Natural, Conditional, and Prescribed Fire Management Plan, 1979," Jim Benedict Collection, Box 5b 1968–1990, Folder Fire Management Plan 1979, Yosemite Park Archives.
18. "Determination of Impact of Yosemite Fire Management Plan, Revised 1/90, M. Finley, Superintendent, Yosemite National Park, and Stanley Albright, Regional Director, Western Region," Jim Benedict Collection, Fire Management Plan, Yosemite Park Archives, 9.
19. For an excellent summary from the perspective of those involved intimately with the NFP's trade-offs, see Steve Botti and Tom Nichols, "National Park Service Fire Restoration, Policies Versus Results: *What*

Went Wrong," Parks Stewardship Forum 37, no. 2 (2021): 353–67, https://doi.org/10.5070/P537253241.

20. Yosemite National Park, *Final Yosemite Fire Management Plan: Environmental Impact Statement* (Yosemite, Calif.: National Park Service, 2004).
21. Memorandum, "Big Meadow Prescribed Fire Review," November 9, 2009, National Park Service, Pacific West Regional Office, Oakland, Calif., https://www.nps.gov/yose/learn/nature/bigmeadowfirefaq.htm.
22. Botti and Nichols, "National Park Service Fire Restoration," 359–60.
23. "Fire Program Review, Yosemite National Park, 5/13/2011," draft copy courtesy Yosemite Fire Management Program.
24. "Big Meadow Prescribed Fire Review, National Park Service, Pacific West Region," November 9, 2009, https://www.nps.gov/yose/learn/nature/bigmeadowfirefaq.htm. The stalling of the program comes from conversations with Dan Buckley and Kelly Martin.
25. Records for the major fires are contained in Fire Management Records, 1930–2020, YOSE 232940, Series III: Wildland Fire Reports, 1970–2001, Subseries C: Major Fire Incidents, 1968–1999, Subseries 1: Canyon Fire, 1968; Subsubseries 2: Buena Vista Fire, 1981; Subseries 3: August Heat Complex, 1987; Subsubseries 4: Walker Fire, 1988, Yosemite Park Archives, El Portal, Calif.
26. Fire Management Records, 1930–2020, YOSE 232940, Series III: Wildland Fire Reports, 1970–2001, Subseries C: Major Fire Incidents, 1968–1999, Subseries 6: Ill fire, 1991; Subseries 7: Horizon Fire, 1994; Subsubseries 8: Ackerson Complex, 1996; Subseries 9: Yosemite Complex, 1999, Yosemite Park Archives, El Portal, Calif.
27. Stephens quoted in Kara Manke, "How Wildfire Restored a Yosemite Watershed," *Berkeley News*, August 9, 2021, https://news.berkeley.edu/2021/08/09/how-wildfire-restored-a-yosemite-watershed/.
28. Scott Stephens et al., "Fire, Water, and Biodiversity in the Sierra Nevada: A Possible Triple Win," *Environmental Research Communications* 3 (2021): 081004, https://doi.org/10.1088/2515-7620/ac17e2.
29. See Richard A. Minnich and Ernesto Franco Vizcaino, *Land of Chamise and Pines: Historical Accounts and Current Status of Northern Baja California's Vegetation*, University of California Publications in Botany 80 (Berkeley: University of California Press, 1998). On the comparison between Illilouette and the Sierra de San Martír, see Scott L. Stephens, Carl N. Skinner, and Samantha J. Gill, "Dendrochronology-Based Fire History of Jeffrey Pine-Mixed Conifer Forests in the Sierra San Pedro Martir, Mexico," *Canadian Journal of Forest Research* 33,

no. 6 (June 2003): 1090–101. On Stephen's research, see Rachelle Hedges and Gabrielle Boisramé, "Illilouette Creek Basin Research and Publications," Stephens Lab, Berkeley Forests, and Center for Fire Research and Outreach, https://nature.berkeley.edu/stephenslab/illiouette-research/.

30. For a good distillation of two decades of research, see Stephens et al., "Fire, Water, and Biodiversity." The roughly eight-year cycle is what Wagener, "Past Fire Incidence," found in his 1961 summary of previous chronologies. More recent studies at Yosemite have concluded with a median fire-return interval of 10 years and noted that the "spatial pattern of burns exhibited self-organizing behavior" (Scholl and Taylor, "Fire Regimes," 362).
31. Park statistics supplied by Kent van Wagtendonk, Yosemite GIS specialist.
32. For a snapshot of the fire, see https://inciweb.nwcg.gov/incident/7674/. On Moore's memo, see Memorandum, "Chief's Wildland Fire Direction," August 2, 2021, U.S. Forest Service, Washington Office, Washington, D.C., https://www.gov.ca.gov/wp-content/uploads/2021/08/8.2.21-USDA-letter.pdf.
33. I take as the failure rate for suppression those fires that escape initial attack (2–3 percent). The Forest Service claim comes from the chief forester's office, in the letter that announced the 90-day moratorium on prescribed fires (see note 35). The Park Service figure comes from Dan Buckley.
34. Statistics from Jan van Wagtendonk, "A Quarter Century of Burning in the Illilouette Creek Watershed," a PowerPoint presentation given in printed form to the 2021 trekkers. On the patterning of these historic fires, see Brandon M. Collins et al., "Spatial Patterns of Large Natural Fires in Sierra Nevada Wilderness Areas," *Landscape Ecology* 22 (2007): 545–57.
35. Will Sarvis, "An Interview with Henry W. Debruin by Telephone from Columbia, Missouri to Arnold, Maryland, 09 July 1998," Oral History Program, State Historical Society of Missouri, 9.
36. Quotes are from Jim Sullivan, fuels technician, to Dave Allen, fire officer in charge of prescribed natural fires, in Fire Management Records, 1930–2010, YOSE 1017, Series III: Wildland Fire Reports, 1970–2001, Subseries C, Subsubseries 7: Horizon Fire, 1994, Yosemite Park Archives, El Portal, Calif. I'm grateful to Robert Reese for explaining the context of the exchange in an email on April 12, 2022.

37. See "NWCG Guidance on Minimum Impact Suppression Tactics in Response to the 10-Year Implementation Plan for Reducing Wildland Fire Risks to Communities and the Environment" available at http://www.wildfirelessons.net/HigherLogic/System/DownloadDocumentFile.ashx?DocumentFileKey=aa4c92ae-ba1f-4e9c-b0b7-b57f55c9a626.
38. Van Wagtendonk, interview, 2011.

Chapter 7

1. Statistics from CalFire at https://www.fire.ca.gov/stats-events/; Bettina Boxall, "Mojave Desert Fire in August Destroyed the Heart of a Beloved Joshua Tree Forest," *Los Angeles Times*, September 6, 2020, https://www.latimes.com/environment/story/2020-09-06/mojave-desert-fire-destroys-the-heart-of-a-beloved-joshua-tree-forest; Jack Herrera, "'Mind-Blowing': Tenth of World's Giant Sequoias May Have Been Destroyed by a Single Fire," *Guardian*, June 3, 2022, https://www.theguardian.com/us-news/2021/jun/02/sequoias-destroyed-california-castle-fire; and Lauren Sommer, "A Single Fire Killed Thousands of Sequoias. Scientists Are Racing to Save the Rest," NPR, September 17, 2021, https://www.npr.org/2021/09/17/1037914390/giant-sequoia-national-park-wildfire-climate-change.
2. Susannah Meadows, "What I Saw in Yosemite Was Devastating," *New York Times*, July 22, 2021, https://www.nytimes.com/2021/07/22/opinion/yosemite-west-coast-smoke.html.
3. Bertrand Russell, *The Scientific Outlook* (New York: Norton, 1931), 140.
4. Quotes from an interview with Dan Buckley, September 10, 2021, a few days before the trek.
5. The fires are chronicled in InciWeb, but a good summary on the day at issue is available in Bill Gabbert, "Large Firing Operations Underway on KNP Complex of Fire," *Wildfire Today*, October 4, 2021, https://wildfiretoday.com/2021/10/04/large-firing-operations-underway-on-knp-complex-of-fires/; Bill Gabbert, "Falling Tree on KNP Complex Injures Four Firefighters," *Wildfire Today*, October 8, 2021, https://wildfiretoday.com/2021/10/08/falling-tree-on-knp-complex-injures-four-firefighters/; and Bill Gabbert, "Officials Estimate Hundreds of Giant Sequoias Were Killed in Windy Fire," *Wildfire Today*, November 4, 2021, https://wildfiretoday.com/2021/11/04/officials-estimate-hundreds-of-giant-sequoias-were-killed-in-the-windy-fire/. See

the National Park Service report on Sequoia-Kings Canyon mortality, "2021 Fire Season Impacts to Giant Sequoias," National Park Service, https://www.nps.gov/articles/000/2021-fire-season-impacts-to-giant-sequoias.htm, accessed November 22, 2021.

6. Botti and Nichols, "National Park Service Fire Restoration." Caprio comment made during an interview with the author in May 2011.
7. Nichols quotes from an email to the author on December 7, 2021.
8. Buckley comments from an email to the author on May 9, 2022.

Epilogue

1. Yosemite Fire and Aviation Information, "Fire Update October 2nd, 2021," https://www.nps.gov/yose/blogs/fire-update-october-2-2021.htm.

Codicil

1. See the InciWeb report (https://inciweb.nwcg.gov/incident/8209) and an update with good maps at Bill Gabbert, "Washburn Fire Grows in Yosemite National Park," *Wildfire Today*, July 10, 2021, https://wildfiretoday.com/2022/07/10/washburn-fire-grows-in-yosemite-national-park/. The fire started after I had received the copyedited book manuscript, but the events were too pertinent to ignore. Since the fire was certain to continue beyond the time I was obligated to return the edits, I decided on a codicil, but that, too, leaves the fire outpacing the book. I chose July 16 as a close date, assuming that the fire was most likely to burn outside the park, and so outside the scope of my subject. For the InciWeb report, see https://inciweb.nwcg.gov/incident/article/8209/69673/. For an update with useful maps, see Bill Gabbert, "The Remaining Battle on Yosemite National Park's Washburn Fire Is on the East Side," *Wildfire Today*, July 15, 2021, https://wildfiretoday.com/2022/07/15/the-remaining-battle-on-yosemite-national-parks-washburn-fire-is-on-the-east-side/. Data on the final size from https://inciweb.nwcg.gov/incident/8209/.
2. Data from inciweb: https://inciweb.nwcg.gov/incident/maps/8332/.
3. On the lawsuit, see "Logging Project in Yosemite National Park Halted After Environmental Lawsuit," *Los Angeles Times*, July 7, 2022, https://www.latimes.com/california/story/2022-07-07/logging-project-yosemite-national-park-halted-after-environmental-lawsuit. Clearly, it was the simple act of cutting that prompted the suit, not the conjured justifications formally submitted.

4. Quote from Ashley Harrell, "Yosemite's Mariposa Grove Will Survive Washburn Fire, Says Park's Forest Ecologist," *SFGATE*, July 11, 2022, https://www.sfgate.com/california-wildfires/article/mariposa-grove-will-survive-fire-17298114.php. For a marvelous deeper discussion, pivoted on Garrett Dickman, see Kyle Dickman, "To Save Sequoias We Must Save Them from Ourselves," *Outside*, July 13, 2022, https://www.outsideonline.com/outdoor-adventure/environment/sequoias-wildfire-california/.

CREDITS

Figure 1. Courtesy Yosemite Park Archives.
Figure 2. Courtesy USGS Photo Library.
Figure 3. From J. B. Leiberg, Forest Conditions in the Northern Sierra Nevada, California (Washington, D.C.: Government Printing Office, 1902).
Figure 4. Courtesy Yosemite Park Archives.
Figure 5. Courtesy Yosemite Park Archives.
Figure 6. Courtesy USGS Photo Library.
Figure 7. Source: U.S. Forest Service Historic Photo Collection, National Archives.
Figure 8. Source: U.S. Forest Service Historic Photo Collection, National Archives.
Figure 9. NPGallery (https://npgallery.nps.gov).
Figure 10. Courtesy Yosemite Park Archives.
Figure 11. Courtesy Yosemite Park Archives.
Figure 12. Courtesy Yosemite Park Archives.
Figure 13. Courtesy Yosemite Park Archives.
Figure 14. Courtesy Yosemite Park Archives.
Figure 15. Courtesy Yosemite Park Archives.
Figure 16. With permission of Aldo Leopold Foundation.
Figure 17. Photo by Bruce Kilgore, NPS.
Figure 18. Photo by George Briggs, NPS.
Figure 19. Courtesy Yosemite Park Archives.
Figure 20. Photo by Dan Taylor, NPS.
Figure 21. Photo courtesy of Dan Buckley, NPS.
Figure 22. Courtesy of Brandon Collins.

Figure 23. Photo by Steve Pyne.
Figure 24. Photo courtesy of NPS.
Figure 25. Original color maps by Jan van Wagtendonk, redone in black and white by Kent van Wagtendonk, NPS.
Figure 26. Photo by Steve Pyne.
Figure 27. Photo courtesy of NPS.
Figure 28. Graphic by author; data courtesy CalFire.
Figure 29. Photo courtesy Chad Anderson, NPS.
Figure 30. Photo by Steve Pyne.
Figure 31. Graphic by author; data from Kent van Wagtendonk, NPS.
Figure 32. Graphic by author; data from Kent van Wagtendonk, NPS.
Map 1. Courtesy Kent van Wagtendonk, NPS.
Map 2. Courtesy Kent van Wagtendonk, NPS.
Map 3. Courtesy Kent van Wagtendonk, NPS.
Map 4. Courtesy Kent van Wagtendonk, NPS.

INDEX

ABOUT THE AUTHOR

Stephen J. Pyne is a fire historian, urban farmer, and emeritus professor at Arizona State University. He spent 15 seasons with the North Rim Longshots, a fire crew at Grand Canyon National Park. He has written fire histories for America, Australia, Canada, Europe (including Russia), and the Earth. A resurvey of the American scene includes *Between Two Fires: A Fire History of Contemporary America* and To the Last Smoke, a nine-volume series of regional reconnaissances.